Chemistry Reference Tables Workbook

2nd Edition

Authors:
Ron Pasto – Retired Chemistry Teacher
William Docekal – Retired Science Teacher

About This Workbook –

Many questions on the New York State Physical Setting/CHEMISTRY Regents Exam may be answered simply by using information given on the Reference Tables. Other questions may require information from the Reference Tables to set up calculations in order to determine the answer. Knowing what information is on the Reference Tables and where to find it are very important steps towards being successful on the Regents exam.

© 2012, Topical Review Book Company, Inc. All rights reserved.
P. O. Box 328
Onsted, MI 49265-0328
www.topicalrbc.com

Chemistry Reference Tables Workbook

The Introduction – Overview, The Chart, and Additional Information –
In these sections, you will find an explanation of the information given on that table. Read each section carefully to fully understand the information given on that table.

Set 1 – Questions and Answers –

After careful reading of the Introduction, Set 1 questions will test your understanding of that particular table. Do all questions in Set 1, and then correct your work by going to the answers for Set 1, which are at the end of the section. The explanation given will help you to understand any mistakes you may have made. If you need additional explanation, ask your teacher for help.

Set 2 – Questions –

The answers to these questions are in a separate answer key. Correctly answering these questions will show yourself and your teacher that you understand the subject matter for that particular table.

All of us at Topical Review Book Company hope that by gaining a complete understanding of the Chemistry Reference Tables, it will help you to increase your knowledge of chemistry and that your grades will improve.

Physical Setting/Chemistry
Reference Tables Workbook
Table of Contents

Name	Value	Unit
Standard Pressure	101.3 kPa 1 atm	kilopascal atmosphere
Standard Temperature	273 K 0°C	kelvin degree Celsius

Overview:

A gas is the form of matter that has no definite shape or volume. Generally, shape is not critical in discussing the properties of a gas, but volume is. The volume of a given mass of gas depends upon the temperature and pressure of the gas.

The Table:

Standard temperature and pressure (STP) are selected conditions that are used in listing and discussing many properties and behavior of gases. The value of standard pressure is given in kilopascals (kPa) and atmospheres (atm). The average atmospheric pressure at sea level and 0°C is one atmosphere (1 atm). Standard temperature is given in Kelvin (K) and degrees Celsius (C). The word degree and its symbol are not used when expressing temperature on the Kelvin scale.

Additional Information:

- Pressure is defined as force per unit area. The pascal (Pa) is a unit of pressure in the Metric System. Since standard atmospheric pressure is 101,300 pascals, it is more conveniently expressed as 101.3×10^3 Pa or 101.3 kilopascals (kPa). See Table C, 2nd bullet.

- The volume of one mole of an ideal gas (molar volume) at STP is 22.4 L.

- In problems dealing with the Combined Gas Law, found in Table T, the initial or final conditions may be given as STP conditions. Use the values given in Table A for standard temperature and pressure.

> Questions dealing with Table A will be given in the **Combined Gas Law, Temperature, and Density** sections of Table T – Important Formulas and Equations.

Heat of Fusion	334 J/g
Heat of Vaporization	2260 J/g
Specific Heat Capacity of $H_2O(\ell)$	4.18 J/g•K

Overview:

Water is a very important substance, not only in chemistry, but for everyday life. Water is also a unique substance in that it readily exists in all 3 phases of matter. When water undergoes a change of phase, energy is either absorbed or released. Three very important physical constants for water are given in this table that show how much heat is involved in phase changes and the change of temperature of water.

The Table:

The Heat of Fusion (H_f) is the amount of heat required to change one gram of ice at 0°C to water at 0°C. Fusion is a term that means melting (change from a solid to a liquid). The Heat of Vaporization (H_v) is the amount of heat required to change one gram of water at 100°C to vapor at 100°C. Vaporization refers to the change from the liquid to the gas phase at the normal boiling point of the liquid. When a phase change occurs, there is no change in temperature. Thus, the units for the Heat of Fusion and the Heat of Vaporization do not have a temperature unit in them. The Specific Heat Capacity (C) is the amount of heat required to change the temperature of one gram of water by 1 kelvin degree. These three heat constants are used in the heat equations on Table T.

Additional Information:

- The joule (J) is the unit for the quantity of heat in the Metric System. See Table D.
  K is the unit for kelvin, a measurement of temperature. See Table D.

- Water has the largest specific heat capacity of all common substances.

Questions dealing with Table B will be given in the **Heat** section of Table T – Important Formulas and Equations.

Table C — Selected Prefixes

Factor	Prefix	Symbol
10^3	kilo-	k
10^{-1}	deci-	d
10^{-2}	centi-	c
10^{-3}	milli-	m
10^{-6}	micro-	μ
10^{-9}	nano-	n
10^{-12}	pico-	p

Overview:

In the sciences, especially chemistry and physics, one uses many numbers that are very small or very large. The system of measurement used in the sciences is the Metric System. In addition to the defined standards for the various quantities, amounts that are multiples or submultiples of 10 are used. Numerical prefixes are frequently used to express multiples of these standards when they are very small or very large.

The Table:

This table lists various multiples of 10 – the Factor (expressed in exponential form), the Prefix and the Symbol for that prefix. By knowing the exponent for a certain prefix, as shown in the Factor column, one can determine how many places the decimal must be moved when performing conversions. For example, as shown in the Factor column, changing from kilo- to centi- (and the reverse) requires a five place decimal move. For most everyday activities, the most common prefixes used are kilo-, centi-, and milli-.

Additional Information:

- The defined units in the Metric System are those of length (the meter – m), mass (the kilogram – kg), and time (the second – s).

- Standard pressure in decimal form is 101,300 Pa (pascal), which can be expressed as 101.3×10^3 Pa. Using the prefix from Table C for 10^3 (kilo), this becomes 101.3 kPa, which is shown in Table A.

- The capacity of memory chips for electronic devices (computers, Ipods, digital cameras) are expressed in megabytes (10^6 or one million bytes) or gigabytes (10^9 or one billion bytes).

- The speed at which a computer completes an operation is expressed in nanoseconds (ns), which are billionths of a second.

Fill in the blanks

1. 1 kilometer = _____ m

2. 250 meters = _____ km

3. 0.33 kilometer = _____ m

4. 1,000 grams = _____ kg

5. 0.0089 kilograms = _____ g

6. 2560 grams = _____ kg

7. 1 centimeter = _____ m

8. 77 centimeters = _____ m

9. 0.2 meters = _____ cm

10. 3.25 meters = _____ cm

11. 0.001 liters = _____ mL

12. 355 milliliters = _____ L

13. 89.4 grams = _____ kg

14. 3,400 milliliters = _____ L

15. 780 milligrams = _____ g

16. 0.0067 grams = _____ mg

17. 1 nanosecond = _____ s

Selected Prefixes

18. Which quantity is equivalent to
    50 kilocalories?

    (1) 5,000 cal
    (2) 0.05 cal
    (3) 5×10^3 cal
    (4) 5×10^4 cal   18 ____

20. Which of the following prefixes is
    used to express 1/10?

    (1) kilo   (3) centi
    (2) deci   (4) milli   20 ____

21. 2,000 mL can be expressed as

    (1) 20 dL   (3) 2 kL
    (2) 2 L   (4) 0.002 kL   21 ____

19. Which of the following prefixes is
    used to express 1/1000?

    (1) kilo
    (2) deci
    (3) centi
    (4) milli   19 ____

22. Which quantity of heat is equal to
    200. joules?

    (1) 20.0 kJ   (3) 0.200 kJ
    (2) 2.00 kJ   (4) 0.0200 kJ   22 ____

23. 9.3×10^{-6} joules is how many microjoules? _____ μJ

24. 5.67 kiloliters is equal to how many liters? _____ L

25. 2.37 kilometers is equal to how many decimeters? _____ dm

26. 4,500 milligrams is equal to how many grams? _____ g

27. 0.023 microseconds is equal to how many nanoseconds? _____ ns

28. 89,600 grams is equal to how many kilograms? _____ kg

Table C – Selected Prefixes
Answers
Set 1

1. 1 kilometer = **1,000** m

2. 250 meters = **0.25** km

3. 0.33 kilometer = **330** m

4. 1,000 grams = **1** kg

5. 0.0089 kilograms = **8.9** g

6. 2,560 grams = **2.56** kg

7. 1 centimeter = **0.01** m

8. 77 centimeters = **.77** m

9. 0.2 meters = **20** cm

10. 3.25 meters = **325** cm

11. 0.001 liters = **1** mL

12. 355 milliliters = **0.355** L

13. 89.4 grams = **0.0894** kg

14. 3,400 milliliters = **3.4** L

15. 780 milligrams = **0.78** g

16. 0.0067 grams = **6.7** mg

17. 1 nanosecond = $\mathbf{1 \times 10^{-9}}$ s or **0.000000001** s

Symbol	Name	Quantity
m	meter	length
g	gram	mass
Pa	pascal	pressure
K	kelvin	temperature
mol	mole	amount of substance
J	joule	energy, work, quantity of heat
s	second	time
min	minute	time
h	hour	time
d	day	time
y	year	time
L	liter	volume
ppm	parts per million	concentration
M	molarity	solution concentration
u	atomic mass unit	atomic mass

Overview:

This table shows units that are commonly used in chemistry. The table shows the Symbol, Name and Quantity being measured.

The Table:

Since the Metric System of measurement is used in chemistry, most of the units on this table are Metric units. Parts per million (ppm) is not a defined unit in the Metric System. For most uses in high school chemistry, the fundamental units used to define the Metric System are those of length, mass, time, temperature, and amount of substance. All other units can be defined in terms of these units. These units will be used throughout this workbook.

Additional Information:

- The Kelvin temperature scale is also known as the absolute temperature scale. Absolute zero (the temperature where molecular motion ceases) is 0 K. This is the lowest possible temperature.

- The mole (mol) is the measure of amount of substance. One mole of a substance is defined as 6.02×10^{23} particles (atoms, molecules or ions) of that substance. This number is referred to as the Avogadro number. One can have a mole of anything, as long as you have the Avogadro number of "things".

- 22.4 L of a gas at STP (one mol) contains 6.02×10^{23} particles of that gas.

- The atomic mass unit, designated by the symbol u, is the unit of mass on an atomic or molecular scale. It is defined as precisely 1/12 the mass of a carbon-12 atom.

Questions and the proper uses of these units will be given within other tables in this workbook.

Table E — Selected Polyatomic Ions

Formula	Name	Formula	Name
H_3O^+	hydronium	CrO_4^{2-}	chromate
Hg_2^{2+}	mercury(I)	$Cr_2O_7^{2-}$	dichromate
NH_4^+	ammonium	MnO_4^-	permanganate
$C_2H_3O_2^-$ CH_3COO^- } acetate	acetate	NO_2^-	nitrite
		NO_3^-	nitrate
CN^-	cyanide	O_2^{2-}	peroxide
CO_3^{2-}	carbonate	OH^-	hydroxide
HCO_3^-	hydrogen carbonate	PO_4^{3-}	phosphate
$C_2O_4^{2-}$	oxalate	SCN^-	thiocyanate
ClO^-	hypochlorite	SO_3^{2-}	sulfite
ClO_2^-	chlorite	SO_4^{2-}	sulfate
ClO_3^-	chlorate	HSO_4^-	hydrogen sulfate
ClO_4^-	perchlorate	$S_2O_3^{2-}$	thiosulfate

Overview:

Polyatomic ions are charged particles composed of two or more atoms. Most are composed of nonmetallic atoms combined with oxygen. Polyatomic ions are quite stable. Thus, in many chemical reactions, the polyatomic ion remains intact and is therefore written the same on the reactant and product side of a chemical equation. These ions are the negative component of many common compounds.

The Table:

The Formula, Name and charge of the ion are given in this table. This information is used in determining the name of a given compound or in writing the formula for a given compound involving a polyatomic ion. The name of these ions usually indicates the element present other than oxygen. For example, the polyatomic ion CrO_4^{2-} is chromate, containing the metal chromium.

Copyright © 2012
Topical Review Book Company

Additional Information:

- The formula for a compound must represent a neutral group of atoms. The total positive oxidation state or number and the total negative oxidation state or number must be equal, leaving a net charge of zero for the group of atoms.

  Example 1: What is the formula for sodium sulfate?
  Answer: Na_2SO_4
  Explanation: From the Periodic Table, the oxidation number of Na is +1 (Na^+) and from Table E, sulfate is SO_4^{2-}. The correct formula must be Na_2SO_4.

  Example 2: What is the formula for calcium phosphate?
  Answer: $Ca_3(PO_4)_2$
  Explanation: From the Periodic Table the oxidation number of Ca is +2 (Ca^{2+}) and from Table E, phosphate is PO_4^{3-}. The correct formula must be $Ca_3(PO_4)_2$.

- In most compounds involving a polyatomic ion, there exist both covalent and ionic bonds. For example, in the compound Na_3PO_4, the phosphate ion (PO_4^{3-}) is formed by the covalent bonding between the nonmetals P and O, while the sodium ion (Na^+) and the phosphate ion are held together by an ionic bond.

- Most of the ions on this table are derived from ternary or oxo acids (acids composed of three elements, two of which are hydrogen and oxygen). If the acid name ends in –ic, it is modified to end in –ate to name the ion. Hence, sulfuric acid (H_2SO_4) gives rise to the sulfate ion (SO_4^{2-}). If the acid name ends in –ous, it is modified to end in –ite to name the ion. Sulfurous acid (H_2SO_3) therefore gives rise to the sulfite ion (SO_3^{2-}).

- In compounds between metals and nonmetals in which the metal shows more than one positive oxidation state, the *Stock System of Nomenclature* is used. In this system, the oxidation state of the metal in the compound is placed in Roman numerals in parentheses immediately following the name of the metal. For example, $CuSO_4$ is copper(II) sulfate since copper has a +2 oxidation state (Cu^{2+}) in this compound.

Specific uses of compounds containing a polyatomic ion:

- Sodium hydrogen carbonate ($NaHCO_3$), also known as sodium bicarbonate, is baking soda or bicarbonate of soda.

- Sodium hypochlorite ($NaClO$) is the substance in Clorox type bleaching solutions and the chlorinating agent used for swimming pools.

- Sodium nitrite ($NaNO_2$) is used in prepared meats to preserve freshness and color.

- Sodium sulfite (Na_2SO_3) is used in salad bars to preserve the freshness of greens.

- Phosphates (PO_4^{3-}) are found in soft drinks.

- Ammonium nitrate (NH_4NO_3) is used in agriculture as a high nitrogen fertilizer.

1. Which polyatomic ion contains the greatest number of oxygen atoms?

   (1) acetate   (3) hydroxide
   (2) carbonate (4) peroxide   1 ____

2. Which formula represents a hydronium ion?

   (1) H_3O^+ (3) OH^-
   (2) NH_4^+ (4) HCO_3^- 2 ____

3. What is the name of the polyatomic ion in the compound Na_2O_2?

   (1) hydroxide   (3) oxide
   (2) oxalate   (4) peroxide   3 ____

4. The name of the compound $KClO_2$ is potassium

   (1) hypochlorite (3) chlorate
   (2) chlorite   (4) perchlorate   4 ____

5. Which element is found in both potassium chlorate and zinc nitrate?

   (1) hydrogen (3) potassium
   (2) oxygen   (4) zinc   5 ____

6. What is the chemical formula for sodium sulfate?

   (1) Na_2SO_3 (3) $NaSO_3$
   (2) Na_2SO_4 (4) $NaSO_4$ 6 ____

7. What is the oxidation state of sodium in $NaNO_2$?

   (1) +1   (3) +2
   (2) –1   (4) –2   7 ____

8. What is the chemical formula for copper(II) hydroxide?

   (1) $CuOH$ (3) $Cu_2(OH)$
   (2) $CuOH_2$ (4) $Cu(OH)_2$ 8 ____

Base your answers to question 9 using the passage below and your knowledge of chemistry.

Acid rain lowers the pH in ponds and lakes and over time can cause the death of some aquatic life. Acid rain is caused in large part by the burning of fossil fuels in power plants and by gasoline-powered vehicles. The acids commonly associated with acid rain are sulfurous acid, sulfuric acid, and nitric acid.

9. *a)* Write the chemical formula of a negative ion present in an aqueous nitric acid solution. ____

   b) Write the chemical formula of a negative ion present in aqueous sulfurous acid. ____

10. A 2.0-liter aqueous solution contains a total of 3.0 moles of dissolved NH_4Cl at 25°C and standard pressure.

   Identify the two ions present in the solute. ____

Selected Polyatomic Ions

11. Which formula is correct for ammonium sulfate?

    (1) NH_4SO_4 (3) $NH_4(SO_4)_2$

    (2) $(NH_4)_2SO_4$ (4) $(NH_4)_2(SO_4)_2$ 11 ____

12. Which formula represents a nitrate ion?

    (1) NO^- (3) NO_3^-

    (2) NO_2^- (4) NO_4^- 12 __3__

13. Which formula represents lead(II) phosphate?

    (1) $PbPO_4$ (3) $Pb_3(PO_4)_2$

    (2) Pb_4PO_4 (4) $Pb_2(PO_4)_3$ 13 ____

14. Which compound has both ionic and covalent bonds?

    (1) CO_2 (3) NaI

    (2) CH_3OH (4) Na_2CO_3 14 __4__

15. The name of the compound $Na_2S_2O_3$ is

    (1) potassium sulfate
    (2) potassium thiosulfate
    (3) sodium sulfate
    (4) sodium thiosulfate   15 __4__

16. What is the oxidation state of Ca in $CaSO_4$?

    (1) –1   (3) –2
    (2) +1   (4) +2   16 __2__

17. Which formula represents strontium phosphate?

    (1) $SrPO_4$ (3) $Sr_2(PO_4)_3$
    (2) Sr_3PO_8 (4) $Sr_3(PO_4)_2$ 17 __4__

18. The chemical bonding in sodium phosphate, Na_3PO_4, is classified as

    (1) ionic, only
    (2) metallic, only
    (3) both covalent and ionic
    (4) both covalent and metallic   18 __3__

19. An antacid contains the acid-neutralizing agent sodium hydrogen carbonate. Write the chemical formula for sodium hydrogen carbonate. _____

20. Write the correct formula for hydrogen sulfate. _____

21. In terms of their formulas, what is the difference between the chlorate and perchlorate ions? _____

22. Write the chemical formula for the most abundant negative ion in an aqueous sodium phosphate solution. _____

23. Write the chemical formula for copper(I) sulfite. _____

1. 2   From Table E, the carbonate ion has the formula CO_3^{2-}. The formula shows three oxygen atoms are present in this ion. The other choices have one or two oxygen atoms.

2. 1   The hydrogen ion (a proton) combines with a water molecule to form the hydronium ion. This polyatomic ion is the first one given in Table E.

3. 4   Sodium peroxide, Na_2O_2, contains the peroxide ion, having the formula O_2^{2-}.

4. 2   Using Table E, the chlorite ion is ClO_2^-. Potassium, having an oxidation number +1, would join with the chlorite ion, forming the compound $KClO_2$, potassium chlorite.

5. 2   Using Table E, potassium chlorate is $KClO_3$, and zinc nitrate is $Zn(NO_3)_2$. Both of these compounds contain oxygen.

6. 2   As shown in the Periodic Table, sodium has an oxidation state of +1 (Na^+). Shown in Table E, the formula for the sulfate ion is SO_4^{2-}. To maintain a neutral compound, two sodium ions are needed to form sodium sulfate, Na_2SO_4.

7. 1   The formula of a compound must represent a neutral group of atoms. The nitrate ion (NO_3^-) has a charge of –1. Thus Na must have an oxidation state of +1 in the formula of $NaNO_3$. This is the oxidation state of sodium as shown in the Periodic Table.

8. 4   In the *Stock System of Nomenclature*, the oxidation state of the metal is shown by Roman numerals. Copper(II) has an oxidation state of +2. From Table E, the hydroxide ion (OH^-) has a charge of –1. For the compound to be neutral, two hydroxide ions are needed. The formula for copper(II) hydroxide would be $Cu(OH)_2$.

9. *a*) NO_3^- If the acid name ends in –ic, it is modified to end in –ate to name the ion. Thus, nitric acid would produce the nitrate ion, NO_3^-.

    b) SO_3^{2-} Sulfurous acid produces the sulfite ion. Remember, if the acid name ends in –ous, it is modified to end in –ite to name the ion.

10. Answer: NH_4^+ and Cl^- *or*   ammonium and chloride

    Explanation: When this compound dissolves in water, it dissociates into the polyatomic ion NH_4^+ and the Cl^- ion.

Ions That Form *Soluble* Compounds	Exceptions
Group 1 ions (Li^+, Na^+, etc.)	
ammonium (NH_4^+)	
nitrate (NO_3^-)	
acetate ($C_2H_3O_2^-$ or CH_3COO^-)	
hydrogen carbonate (HCO_3^-)	
chlorate (ClO_3^-)	
halides (Cl^-, Br^-, I^-)	when combined with Ag^+, Pb^{2+}, or Hg_2^{2+}
sulfates (SO_4^{2-})	when combined with Ag^+, Ca^{2+}, Sr^{2+}, Ba^{2+}, or Pb^{2+}

Ions That Form *Insoluble* Compounds*	Exceptions
carbonate (CO_3^{2-})	when combined with Group 1 ions or ammonium (NH_4^+)
chromate (CrO_4^{2-})	when combined with Group 1 ions, Ca^{2+}, Mg^{2+}, or ammonium (NH_4^+)
phosphate (PO_4^{3-})	when combined with Group 1 ions or ammonium (NH_4^+)
sulfide (S^{2-})	when combined with Group 1 ions or ammonium (NH_4^+)
hydroxide (OH^-)	when combined with Group 1 ions, Ca^{2+}, Ba^{2+}, Sr^{2+}, or ammonium (NH_4^+)

*compounds having very low solubility in H_2O

Overview:

This table is used to determine whether a particular compound is soluble or insoluble in water (aqueous) solution. If an insoluble substance is formed in the reaction between two aqueous solutions of different salts (ionic compounds), it is called a precipitate and settles to the bottom of the container.

The Table:

The top chart shows ions that form soluble compounds with some exceptions noted. The bottom chart shows ions that form insoluble or nearly insoluble compounds with some exceptions noted.

Some general rules for solubility can be stated using information from this table:
- all compounds containing Group 1 ions are soluble in water
- all compounds containing ammonium, nitrate, acetate, hydrogen carbonate, and chlorate ions are soluble in water

Be very careful to note the **Exceptions** columns. Many questions on the regents involve these examples.

Additional Information:

- The halides are negative ions formed from Group 17 elements, known as the halogens.

- Since a relatively large amount of a soluble substance may be dissolved in a given amount of water, these solutions may be concentrated (strong solutions).

- Since only a small amount of an insoluble substance dissolves in a given amount of water, these solutions are dilute (weak solutions).

- Soluble ionic substances (salts) dissolved in water form solutions that readily conduct an electric current. They are referred to as strong electrolytes.

- The notation (s) following the formula of a substance indicates that the substance is a solid or insoluble in water (a precipitate). The notation (aq) following a formula indicates an aqueous solution of that substance (soluble in water).

- When an insoluble substance (precipitate) is formed, it may be separated from the rest of the solution by the process of filtration. However, a soluble solute cannot be separated from the solvent by filtration.

1. According to Table F, which of these salts is *least* soluble in water?

   (1) LiCl   (3) $FeCl_2$
   (2) RbCl   (4) $PbCl_2$ 1 _____

2. Which compound is insoluble in water?

   (1) $BaSO_4$ (3) $KClO_3$
   (2) $CaCrO_4$ (4) Na_2S 2 _____

3. Which ion, when combined with chloride ions, Cl^-, forms an insoluble substance in water?

   (1) Fe^{2+} (3) Pb^{2+}
   (2) Mg^{2+} (4) Zn^{2+} 3 _____

4. Based on Reference Table F, which of these saturated solutions has the *lowest* concentration of dissolved ions?

   (1) NaCl(aq)
   (2) $MgCl_2$(aq)
   (3) $NiCl_2$(aq)
   (4) AgCl(aq)   4 _____

5. According to Reference Table F, which of these compounds is most soluble at 298 K and 1 atm?

   (1) AgCl   (3) $MgCrO_4$
   (2) AgI   (4) $PbCO_3$ 5 _____

6. Based on Reference Table F, which salt is the most soluble?

   (1) AgI   (3) $ZnCO_3$
   (2) AgBr   (4) K_2SO_4 6 _____

7. Based on Reference Table F, which compound could form a concentrated solution?

   (1) AgBr   (3) Ag_2CO_3
   (2) AgCl   (4) $AgNO_3$ 7 _____

8. Which compound when stirred in water will not pass through filter paper?

   (1) NaCl   (3) $Mg(OH)_2$
   (2) NH_4S (4) LiCl   8 _____

9. A student observed the following reaction:

   $$AlCl_3(aq) + 3NaOH(aq) \rightarrow Al(OH)_3(s) + 3NaCl(aq)$$

   After the products were filtered, which substance remained on the filter paper?

   (1) NaCl   (3) $AlCl_3$
   (2) NaOH   (4) $Al(OH)_3$ 9 _____

10. Which barium salt is insoluble in water?

    (1) $BaCO_3$ (3) $Ba(ClO_3)_2$
    (2) $BaCl_2$ (4) $Ba(NO_3)_2$ 10 _____

Base your answers to question 11 using the information below and your knowledge of chemistry.

In a laboratory activity, 0.500 mole of NaOH(s) is partially dissolved in distilled water to form 400. milliliters of NaOH(aq). This solution is then used to titrate a solution of HNO_3(aq).

11. *a*) Identify the negative ion produced when the NaOH(s) is dissolved in distilled water.

_____

b) Another student substituted $Mg(OH)_2$ for NaOH to make a solution to be use in this tritration. Which compound would be more soluble?

_____

Base your answers to question 12 using the information below and your knowledge of chemistry.

Calcium hydroxide is commonly known as agricultural lime and is used to adjust the soil pH. Before the lime was added to a field, the soil pH was 5. After the lime was added, the soil underwent a 100-fold decrease in hydronium ion concentration.

12. *a*) According to Reference Table F, is calcium hydroxide soluble in water?

_____

b) Identify another hydroxide compound that contains a Group 2 element and is soluble in water.

_____

13. Give a statement on the solubility of $Pb(C_2H_3O_2)_2$.

It forms Soluble Compounds in water + an ion

14. The dissolving of solid lithium bromide in water is represented by the balanced equation below.

$$LiBr(s) \xrightarrow{H_2O} Li^+(aq) + Br^-(aq)$$

Based on Table F, identify *one* ion that reacts with Br^- ions in an aqueous solution to form an insoluble compound. Ag

Solubility Guidelines

15. Which of the following compounds is *least* soluble in water?

(1) copper(II) chloride
(2) aluminum acetate
(3) iron(III) hydroxide
(4) potassium sulfate   15 _____

16. According to Reference Table F, which of these compounds is the least soluble in water?

(1) K_2CO_3 (3) $Ca_3(PO_4)_2$
(2) $KC_2H_3O_2$ (4) $Ca(NO_3)_2$ 16 _____

Least   Soluble = insoluble

17. Which compound, when stirred in water, will not pass through filter paper?

(1) Hg_2Cl_2 (3) Na_3PO_4
(2) $MgCrO_4$ (4) Na_2S 17 _____

18. Which two solutions, when mixed together, will undergo a double replacement reaction and form a white, solid substance?

(1) NaCl(aq) and $LiNO_3$(aq)
(2) KCl(aq) and $AgNO_3$(aq)
(3) KCL (aq) and LiCl(aq)
(4) $NaNO_3$(aq) and $AgNO_3$(aq)   18 _____

19. A student observed the following reaction:

$$NaCl(aq) + AgNO_3(aq) \rightarrow NaNO_3(aq) + AgCl(s)$$

After the products were filtered, which substance remained on the filter paper?

(1) NaCl   (3) $NaNO_3$
(2) $AgNO_3$ (4) AgCl   19 _____

20. According to Table F, which compound is soluble in water?

(1) barium phosphate
(2) calcium sulfate
(3) silver iodide
(4) magnesium chromate   20 _____

Base your answer to question 21 using the information below and your knowledge of chemistry.

Gypsum is a mineral that is used in the construction industry to make drywall (sheetrock). The chemical formula for this hydrated compound is $CaSO_4 \cdot 2H_2O$. A hydrated compound contains water molecules within its crystalline structure. Gypsum contains 2 moles of water for each 1 mole of calcium sulfate.

21. Give a statement on the solubility of gypsum in water.

It is insoluble because of an exception.

Base your answer to question 22 using the information below and your knowledge of chemistry.

Electroplating is an electrolytic process used to coat metal objects with a more expensive and less reactive metal. The diagram below shows an electroplating cell that includes a battery connected to a silver bar and a metal spoon. The bar and spoon are submerged in $AgNO_3(aq)$.

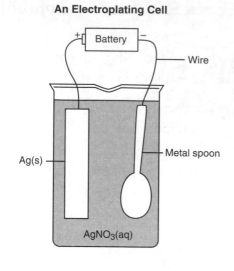

An Electroplating Cell

22. Explain why $AgNO_3$ is a better choice than $AgCl$ for use in this electrolytic process.

_____

_____

23. A 1.0-gram strip of zinc is reacted with hydrochloric acid in a test tube. The unbalanced equation below represents the reaction.

$$Zn(s) + HCl(aq) \rightarrow H_2(g) + ZnCl_2(aq)$$

Explain, using information from Reference Table F, why the symbol (aq) is used to describe the product $ZnCl_2$.

Aq tells us the substance was dissolved in water.

24. Based on Reference Table F, describe the solubility of zinc sulfide in water.

Insoluble

25. Based on Reference Table F, describe the solubility of magnesium hydroxide in water.

In Soluble

26. Solutions of ammonium sulfide and calcium nitrate are allowed to react according to the equation:

$$(NH_4)_2S + Ca(NO_3)_2 \rightarrow 2NH_4NO_3 + CaS$$

If the mixture of products is filtered, identify which product, if any, will remain on the filter paper.

CaS because it is insoluble.

Solubility Guidelines

Set 1

1. 4   In Table F on the left side of the chart, locate "Ions That Form *Soluble* Compounds". Read down to the halides in which Cl^- is shown. To the right of this in the Exceptions column Pb^{2+} is listed. Thus, $PbCl_2$ would be insoluble.

2. 1   On the left side of Table E, in the "Ions That Form *Soluble* Compounds" locate sulfates (SO_4^{2-}). To the right of this are the exceptions, of which Ba^{2+} is listed.

3. 3   Open to Table F. Using the left chart, "Ions That Form *Soluble* Compounds", locate Cl^- in the halide row. To the immediate right are the exceptions, which include compounds containing Pb^{2+}.

4. 4   A solution with the lowest concentration of ions would be one that the solute is insoluble or slightly soluble. As shown in Table F, when Cl^- joins with Ag^+ it is insoluble, being listed in the Exception column.

5. 3   Locate chromate (CrO_4^{2-}) in the chart of "Ions That Form *Insoluble* Compounds". In the column next to it are given the exceptions. When chromate is combined with Mg, it will be soluble.

6. 4   The sulfates are listed as "Ions That Form *Soluble* Compounds". In the Exceptions column potassium (K^+) is not listed, thus K_2SO_4 must be soluble.

7. 4   A concentrated solution would occur with a solute that is very soluble. Of the given compounds, only $AgNO_3$ (see nitrate, NO_3^-), would be soluble, able to make a concentrated solution.

8. 3   A compound that is insoluble in water will remain as a solid and not pass through filter paper. Locate the column, "Ions That Form *Insoluble* Compounds". In this column, go down to the hydroxide row. Magnesium (Mg^{2+}) is not listed in the Exceptions column, therefore $Mg(OH)_2$ would be insoluble.

9. 4   The equation shows that $Al(OH)_3(s)$ is a solid that would have precipitated out of the solution. Being a solid, it will be caught in the filter paper, while the soluble solute, NaCl(aq), will pass through the filter paper.

10. 1   Turn to Table F. The carbonate ion (CO_3^{2-}) is listed in the *Insoluble* Compounds column. Ba^{2+} is not listed as an exception, thus $BaCO_3$ will be insoluble in water.

11. *a)* Answer: OH⁻ *or* hydroxide

Explanation: When NaOH(s) is dissolved in water, it will produce sodium and hydroxide ions. The hydroxide ion is a negative ion (see Table E).

b) Answer: NaOH

Explanation: Using Table F, locate hydroxide (OH⁻) in the "Ions That Form *Insoluble* Compounds" column. To the right of this is given the Exceptions. Sodium (a Group 1 element) is listed as an exception, making NaOH soluble. $Mg(OH)_2$ is insoluble.

12. *a)* Answer: Yes

Explanation: Locate hydroxide (OH⁻) in the left column of "Ions That Form *Insoluble* Compounds". To the right of this is given the Exceptions. Ca^{2+} is listed as an exception, making $Ca(OH)_2$ soluble.

b) Answer: $Ba(OH)_2$ *or* $Sr(OH)_2$

Explanation: Locate hydroxide (OH⁻) in the "Ions That Form *Insoluble* Compounds" column. To the right of this is given the exceptions, which include Ca^{2+}, Ba^{2+}, and Sr^{2+}. These soluble ions are Group 2 ions.

13. Answer: This compound is soluble.

Explanation: In the "Ions That Form *Soluble* Compounds", locate acetate ($C_2H_3O_2^-$). Lead (Pb^{2+}) is not listed in the Exceptions column, thus lead acetate would be soluble.

14. Answer: Ag^+ *or* Hg_2^{2+} *or* Pb^{2+} *or* lead(II) *or* mercury(I)

Explanation: From Table F locate the halides column. The exceptions for the soluble of the Br⁻ ion (making them insoluble) are given in the next column.

Solubility Guidelines

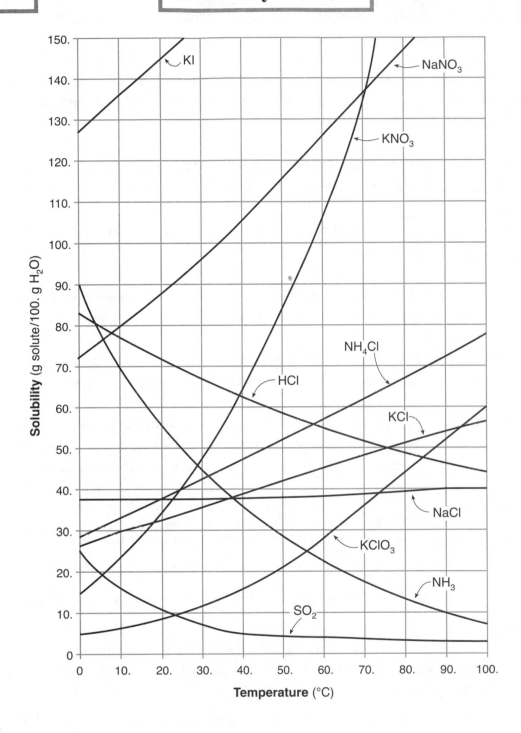

Solubility (g solute/100. g H₂O) vs **Temperature (°C)**

KI, NaNO₃, KNO₃, NH₄Cl, HCl, KCl, NaCl, KClO₃, NH₃, SO₂

Overview:

A solution is a homogeneous mixture of two or more substances. A solution has two components, the dissolved substance, called the solute, and the dissolving substance, called the solvent. In most solutions, the solvent is water and these are called aqueous solutions (aq). Temperature is one factor that determines the solubility of a solute in water. If the solute is a gas, pressure, as well as temperature, determines the solubility of that solute. The solubility of a solute as a function of temperature gives rise to a solubility curve.

The Table:

This table shows the mass of solute (dissolved substance), in grams (g), that can be dissolved in 100. g of H_2O as a function of temperature. From the intersection point of a solubility curve and a given temperature line, reading directly to the left gives the amount of that solute, in g, needed to saturate 100. g of H_2O at that temperature. A saturated solution contains the maximum amount of that solute that can be dissolved in 100. g of H_2O at that temperature. An equilibrium exists between dissolved solute and excess undissolved solute in a saturated solution.

Since the solubility of a solid solute increases with temperature, if the solution is cooled sufficiently, solute may start to drop out or precipitate out of solution. The resulting solution is then a saturated solution.

If the amount of solute dissolved in 100. g of H_2O is below the amount read from the solubility curve at that temperature, the solution is unsaturated.

If the amount of solute dissolved in 100. g of H_2O is greater than that amount read from the solubility curve at that temperature, the solution is supersaturated.

The graph shows that the solubility of solid solutes in H_2O generally increases as the temperature increases, while the solubility of gases (HCl, NH_3 and SO_2) decreases as the temperature increases.

If the amount of H_2O is different from 100. g, the amount of solute needed to saturate that amount of H_2O will change accordingly. For example, if 50. g of H_2O is used, take half the amount of solute as read from the table. If 200. g of H_2O is used, take twice the amount of solute as read from the table.

Additional Information:

- A solution is homogeneous since the solute is distributed uniformly throughout.

- The dissolved solute in a solution cannot be separated from the solvent by filtration.

- The presence of solute raises the boiling point of the solvent and lowers the freezing point of the solvent.

- Pressure has a negligible affect on the solubility of a solid in water. However, an increase in pressure increases the solubility of a gas in water and a decrease in pressure decreases the solubility of a gas in water.

- A supersaturated solution is very unstable. Any disturbance, such as stirring or adding a crystal of the solute, will cause the excess solute to crystallize or drop out of solution, forming a saturated solution.

- In using this table, be sure to use the correct solubility curve and the correct temperature line.

1. A dilute, aqueous potassium nitrate solution is best classified as a

   (1) homogeneous compound
   (2) homogeneous mixture
   (3) heterogeneous compound
   (4) heterogeneous mixture   1 _____

2. According to Reference Table G, which substance forms an unsaturated solution when 80. grams of the substance is added in 100. grams of H_2O at 10°C?

   (1) KI   (3) $NaNO_3$
   (2) KNO_3 (4) NaCl   2 _____

3. A saturated solution of $NaNO_3$ is prepared at 60°C using 100. grams of water. As this solution is cooled to 10°C, $NaNO_3$ precipitates (settles) out of the solution. The resulting solution is saturated. Approximately how many grams of $NaNO_3$ settled out of the original solution?

   (1) 46 g   (3) 85 g
   (2) 61 g   (4) 126 g   3 _____

4. One hundred grams of water is saturated with NH_4Cl at 50°C. According to Table G, if the temperature is lowered to 10°C, what is the total amount of NH_4Cl that will precipitate?

   (1) 5.0 g   (3) 30. g
   (2) 17. g   (4) 50. g   4 _____

5. Based on Reference Table G, what is the maximum number of grams of KCl(s) that will dissolve in 200. grams of water at 50°C to produce a saturated solution?

   (1) 38g   (3) 58 g
   (2) 42 g   (4) 84 g   5 _____

6. According to Reference Table G, which solution is saturated at 30°C?

   (1) 12 grams of $KClO_3$ in 100. grams of water
   (2) 12 grams of $KClO_3$ in 200. grams of water
   (3) 30 grams of NaCl in 100. grams of water
   (4) 30 grams of NaCl in 200. grams of water   6 _____

7. A mixture of crystals of salt and sugar is added to water and stirred until all solids have dissolved. Which statement best describes the resulting mixture?

   (1) The mixture is homogeneous and can be separated by filtration.
   (2) The mixture is homogeneous and cannot be separated by filtration.
   (3) The mixture is heterogeneous and can be separated by filtration.
   (4) The mixture is heterogeneous and cannot be separated by filtration.   7 _____

8. A solution that is at equilibrium must be

   (1) concentrated   (3) saturated
   (2) dilute   (4) unsaturated   8 _____

9. Which formula represents a mixture?

   (1) $C_6H_{12}O_6(\ell)$ (3) LiCl(aq)
   (2) $C_6H_{12}O_6(s)$ (4) LiCl(s)   9 _____

10. What occurs when NaCl(s) is added to water?

    (1) The boiling point of the solution increases, and the freezing point of the solution decreases.
    (2) The boiling point of the solution increases, and the freezing point of the solution increases.
    (3) The boiling point of the solution decreases, and the freezing point of the solution decreases.
    (4) The boiling point of the solution decreases, and the freezing point of the solution increases.

10 _____

11. According to Reference Table G, how many grams of $KClO_3$ must be dissolved in 100. grams of H_2O at 10°C to produce a saturated solution?

_____ g

Base your answers to question 12 on the information below and on your knowledge of chemistry.

    When cola, a type of soda pop, is manufactured, $CO_2(g)$ is dissolved in it.

12. *a*) A capped bottle of cola contains $CO_2(g)$ under high pressure. When the cap is removed, how does pressure affect the solubility of the dissolved $CO_2(g)$?

_____

_____

    b) A glass of cold cola is left to stand 5 minutes at room temperature. How does temperature affect the solubility of the $CO_2(g)$?

_____

_____

    c) In the accompanying space, draw a set of axes and label one of them "Solubility" and the other "Temperature."

    d) Draw a line to indicate the solubility of $CO_2(g)$ versus temperature on the axes drawn in part *c*.

13. Given the data table below showing the solubility of salt **X**:

Temperature (C°)	Mass of Solute per 100. g of H_2O
10	22
25	40
30	48
60	107
70	135

a) Which salt on Table G is most likely to be salt **X**? _____

b) On the graph below, scale and label the *y*-axis including appropriate units.

Solubility of Salt X

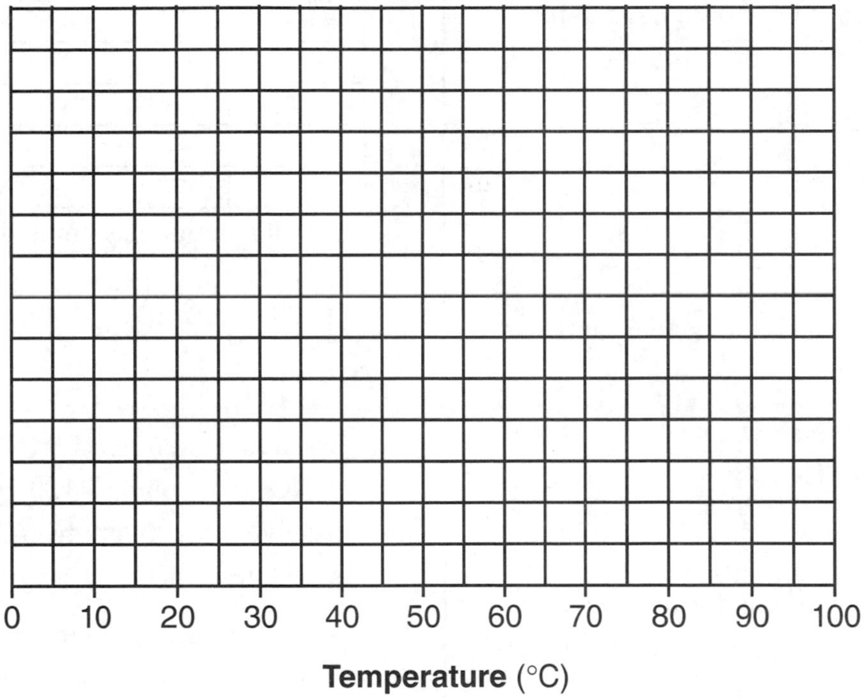

Temperature (°C)

c) Plot the data from the data table. Surround each point with a small circle and draw a best-fit curve for the solubility of salt **X**.

d) Using your graph, predict the solubility of salt **X** at 50°C. _____

e) If the pressure on the salt solution was increased, what affect would this pressure change have on the solubility of the salt?

_____

f) From you graph, how many grams of this solute would dissolve in 200. g of H_2O at 50°C? _____ g

14. According to Reference Table G, which solution at equilibrium contains 50. grams of solute per 100. grams of H_2O at 75°C?

    (1) an unsaturated solution of KCl
    (2) an unsaturated solution of $KClO_3$
    (3) a saturated solution of KCl
    (4) a saturated solution of $KClO_3$ 14 _____

15. A solution contains 35. grams of KNO_3 dissolved in 100. grams of water at 40°C. How much more KNO_3 would have to be added to make it a saturated solution?

    (1) 29 g   (3) 12 g
    (2) 24 g   (4) 4 g   15 _____

16. Which compound forms a saturated solution at 40°C that contains 46. grams per 10. grams of water?

    (1) KNO_3 (3) $NaNO_3$
    (2) NH_4Cl (4) KCl   16 _____

17. An unsaturated solution is formed when 80. grams of a salt is dissolved in 100. grams of water at 40.°C. This salt could be

    (1) KCl   (3) NaCl
    (2) KNO_3 (4) $NaNO_3$ 17 _____

18. Which solution has the lowest freezing point?

    (1) 10. g of KI dissolved in 100. g of water
    (2) 20. g of KI dissolved in 200. g of water
    (3) 30. g of KI dissolved in 100. g of water
    (4) 40. g of KI dissolved in 200. g of water

                                       18 _____

19. At STP, which of these substances is most soluble in H_2O?

    (1) NH_3 (3) HCl
    (2) KCl   (4) $NaNO_3$ 19 _____

20. A student prepares four aqueous solutions, each with a different solute. The mass of each dissolved solute is shown in the table.

Mass of Dissolved Solute for Four Aqueous Solutions

Solution Number	Solute	Mass of Dissolved Solute (per 100. g of H_2O at 20.°C)
1	KI	120. g
2	$NaNO_3$	88 g
3	KCl	25 g
4	$KClO_3$	5 g

Which solution is saturated?

    (1) 1   (3) 3
    (2) 2   (4) 4   20 _____

21. Based on Reference Table G, a solution of $NaNO_3$ that contains 120. grams of solute dissolved in 100. grams of H_2O at 50°C is best described as

    (1) saturated and dilute
    (2) saturated and concentrated
    (3) supersaturated and dilute
    (4) supersaturated and concentrated

                                       21 _____

22. When 5 grams of KCl are dissolved in 50. grams of water at 25°C, the resulting mixture can be described as

    (1) heterogeneous and unsaturated
    (2) heterogeneous and supersaturated
    (3) homogeneous and unsaturated
    (4) homogeneous and supersaturated

                                       22 _____

23. A student adds solid KCl to water in a flask. The flask is sealed with a stopper and thoroughly shaken until no more solid KCl dissolves. Some solid KCl is still visible in the flask. The solution in the flask is

(1) saturated and is at equilibrium with the solid KCl
(2) saturated and is not at equilibrium with the solid KCl
(3) unsaturated and is at equilibrium with the solid KCl
(4) unsaturated and is not at equilibrium with the solid KCl   23 ____

24. Which sample is a homogeneous mixture?

(1) NaCl(s)   (3) NaCl(g)
(2) NaCl(l)   (4) NaCl(aq)   24 ____

25. What is the mass of NH_4Cl that must dissolve in 200. grams of water at 50.°C to make a saturated solution?

(1) 26 g   (3) 84 g
(2) 42 g   (4) 104 g   25 ____

26. Compared to pure water, an aqueous solution of calcium chloride has a
(1) higher boiling point and higher freezing point
(2) higher boiling point and lower freezing point
(3) lower boiling point and higher freezing point
(4) lower boiling point and lower freezing point   26 ____

Base your answers to question 27 using the information below and your knowledge of chemistry.

A student uses 200. grams of water at a temperature of 60°C to prepare a saturated solution of potassium chloride, KCl.

27. a) Identify the solute in this solution. _____

b) According to Reference Table G, how many grams of KCl must be used to create this saturated solution? _____

c) This solution is cooled to 10°C and the excess KCl precipitates (settles out). The resulting solution is saturated at 10°C. How many grams of KCl precipitated out of the original solution?

_____

28. Sulfur dioxide, SO_2, is one of the gases that reacts with water to produce acid rain. According to Reference Table G, describe how the solubility of sulfur dioxide in water is affected by an increase in water temperature.

_____

_____

Base your answers to question 29 using your knowledge of chemistry and the accompanying data table.

The Solubility of the Solute at Various Temperatures

Temperature (°C)	Solute per 100 g of H₂O(g)
0	18
20	20
40	24
60	29
80	36
100	49

29. *a)* On the grid below, mark an appropriate scale on the axis labeled "Solute per 100. g of H_2O(g)." An appropriate scale is one that allows a trend to be seen.

b) On the same grid, plot the data from the data table. Circle and connect the points. Example:

c) Based on the data table, if 15 grams of solute is dissolved in 100. grams of water at 40°C, how many more grams of solute can be dissolved in this solution to make it saturated at 40°C?

_____ g

d) If a saturated solution of this solute at 80°C is cooled down to 30°C, how many grams of the solute would be found on the bottom of the beaker? _____ g

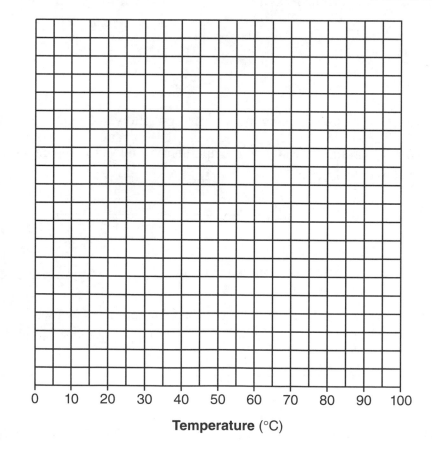

Solute per 100 g of H₂O(g)

Temperature (°C)

30. Identify the type of mixture formed when the NH_4NO_3(s) is completely dissolved in water.

_____

Solubility Curves

1. 2   A mixture is two or more different substances that are not chemically combined. Mixtures may be homogenous or heterogeneous in nature. Solutions are homogenous mixtures in which each substance retains its properties. Remember, a homogeneous mixture has a uniform composition throughout, while a heterogeneous mixture has parts with different compositions.

2. 1   In Table G, go up the 10 degree line until it intersects the 80 gram line. Any solubility line below this point of intersection would produce a saturated solution and any solubility line above this point would produce an unsaturated solution. KI solubility curve is above this point, making an unsaturated solution at this temperature.

3. 1   Open to Table G, follow the 60°C line until in intersects the $NaNO_3$ line. At this intersection point the solution is saturated, holding 126 grams of solute. At 10°C a saturated solution of $NaNO_3$ will contain 80 grams of solute. As the solution cools from 60°C to 10°C, 46 g of $NaNO_3$ will precipitate out of the solution.

4. 2   Referring to Table G, the solubility of NH_4Cl in 100. g of H_2O at 50°C is approximately 52 g. At 10°C, the solubility is approximately 35 g. As the solution cools from 50°C to 10°C, 17 g of NH_4Cl will precipitate out of the solution.

5. 4   At 50°C, a saturated solution of KCl holds 42 g in 100. g water. In 200. g of water at 50°C, twice as much KCl can be dissolved, or 84 g.

6. 1   The graph shows that at the intersection point of 30°C and 12 g is on the $KClO_3$ solubility line. This would make this solution saturated. All other choices are unsaturated solutions.

7. 2   By definition, a solution is a homogeneous mixture of two or more substances. The components of a solution, the solute (dissolved substance) and the solvent (dissolving substance) cannot be separated by the process of filtration.

8. 3   A saturated solution is a solution that has dissolved all the solute possible at a given temperature. In a saturated solution, an equilibrium exists between dissolved and undissolved solute. If more solute is added to a saturated solution, it will settle to the bottom where the rate at which the solid solute dissolves becomes equal to the rate at which the solute crystallizes from the solution.

9. 3   The aqueous symbol (aq) shows that the compound LiCl has dissolved in water, producing a homogenous mixture.

10. 1   The addition of a soluble solute increases the boiling point of the solvent, and at the same time, lowers the freezing point of the solvent.

11.   Answer: correct range is 6 g – 8 g

     Explanation: The saturation point of a solution of $KClO_3$ at 10°C occurs at the intersection of this temperature and the $KClO_3$ solubility curve.

12. *a*) Answer: As pressure decreases, the solubility of a gas in a liquid decreases.

Explanation: When the cap is removed, the pressure decreases, causing a decrease in the solubility of the $CO_2(g)$. This is why bubbles of $CO_2(g)$ are seen rising to the top of the soda in an open bottle.

b) Answer: As the temperature increases, the solubility of a gas in a liquid decreases.

Explanation: As the temperature of the soda increases to room temperature, the solubility of the $CO_2(g)$ decreases, causing the $CO_2(g)$ to come out of solution as bubbles.

c) Answer: 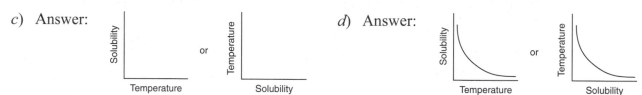   *d*) Answer:

Explanation: As temperature increases the solubility of a gas decreases.

13. *a*) Answer: KNO_3 *or*   potassium nitrate

Explanation: Matching the data in the given data table to the solubility curves found on Table G, the given data identifies the solubility curve as potassium nitrate.

b) Possible labels for the *y*-axis include:

Mass of solute per 100. g of $H_2O(g)$ *or*   grams solute (g)   *or*   mass solute (g)

See graph for proper scaling for *y*-axis.

c) Answer: See graph.

Explanation: Points must be plotted accurately to within ±0.3 of a grid space. The line must be a best-fit curve.

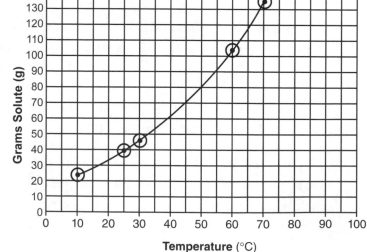

d) Answer: 80 g ±5 g

Explanation: Follow the 50°C line up until it intersects the solubility curve. Read directly to the left to find the solubility of **X**.

e) Answer: No change would occur.

Explanation: Pressure does not affect the solubility of a solid (salt) solute in an aqueous solution.

f) Answer: 160 g

Explanation: In 100. g of H_2O at 50°C, 80 g can be dissolved. In 200. g of H_2O at 50°C, 160 g can be dissolved.

Solubility Curves

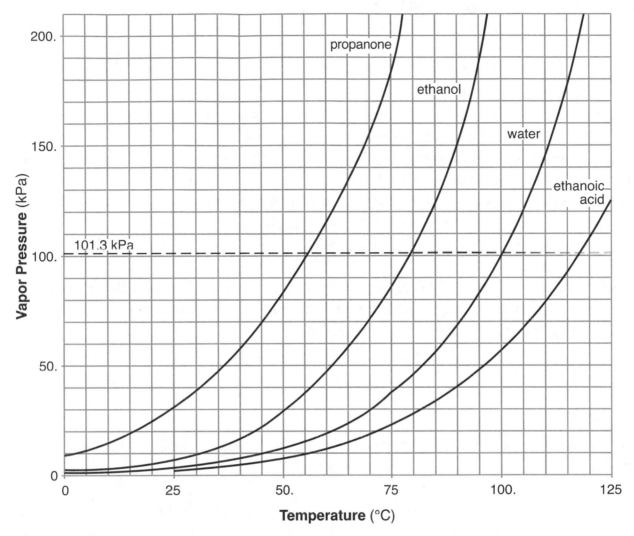

Overview:

A liquid is the form of matter that has definite volume but no definite shape. A liquid takes the shape of the container it is in. Above the surface of a liquid, there is always found the gaseous form of that liquid, called a vapor. The term vapor refers to the gas phase of a substance that is ordinarily a solid or liquid at that temperature. This vapor above the surface of a liquid exerts a characteristic pressure called vapor pressure.

The Table:

This table shows the vapor pressure, in kPa, of four liquids as a function of temperature. The graph shows that propanone has the greatest vapor pressure at any given temperature compared to the other three liquids, while ethanoic acid has the lowest vapor pressure at any given temperature compared to the other three liquids. To determine the vapor pressure of a liquid at a specific temperature, move directly up from the given temperature until you reach the intersection point of the liquid's vapor pressure curve. Reading across to the vapor pressure axis gives the vapor pressure of that liquid at that temperature. The dotted horizontal line labeled 101.3 kPa is standard pressure (see Table A).

Temperature vs. Vapor Pressure

As the temperature increases, the vapor pressure increases. This is due to an increased amount of vapor and the greater average kinetic energy of the vapor particles. As the pressure on the surface of a liquid increases, the boiling point of the liquid increases. This is caused by the need to reach a higher vapor pressure to equal the increased pressure on the surface of the liquid.

Boiling Point and Vapor Pressure

The boiling point of a liquid is the temperature at which the vapor pressure is equal to the atmospheric pressure on the surface of the liquid. Therefore, when a liquid is boiling, the atmospheric pressure on the liquid can be read from the vapor pressure axis since they are equal to each other. When the atmospheric pressure is equal to standard pressure, the boiling point is called the normal boiling point. Reading from the graph at standard pressure (101.3 kPa), the normal boiling points of propanone, ethanol, water and ethanoic acid are 56°C, 79°C, 100°C and 117°C, respectively.

Intermolecular Attraction

A higher boiling point for a liquid indicates a greater attraction between the molecules of that liquid. The vapor pressure curves on Table H indicate that propanone has the weakest intermolecular attraction and ethanoic acid has the greatest intermolecular attraction.

Additional Information:

- The vapor pressure depends only upon the nature of the liquid and the temperature. It does not depend upon the amount of liquid.

- If a temperature-pressure point lies on one of the vapor pressure curves, the liquid is boiling, changing from the liquid to the gas phase. If the intersection point of the temperature and atmospheric pressure (read from the vapor pressure axis) of the substance is to the left of its vapor pressure curve, that substance is a liquid. If the intersection point lies to the right of the vapor pressure curve, it is a gas. For example, at 25°C and 150 kPa pressure, propanone is in the liquid phase, while at 25°C and 20 kPa pressure, propanone is in the gaseous phase.

1. Which substance has the lowest vapor pressure at 75°C?

   (1) water
   (2) ethanoic acid
   (3) propanone
   (4) ethanol   1 _____

2. According to Reference Table H, what is the vapor pressure of propanone at 45°C?

   (1) 22 kPa   (3) 70. kPa
   (2) 33 kPa   (4) 98 kPa   2 _____

3. The boiling point of a liquid is the temperature at which the vapor pressure of the liquid is equal to the pressure on the surface of the liquid. What is the boiling point of propanone if the pressure on its surface is 48 kilopascals?

   (1) 25°C   (3) 35°C
   (2) 30.°C   (4) 40.°C   3 _____

4. At which temperature is the vapor pressure of ethanol equal to the vapor pressure of propanone at 35°C?

   (1) 35°C   (3) 82°C
   (2) 60.°C   (4) 95°C   4 _____

5. As the temperature of a liquid increases, its vapor pressure

   (1) decreases
   (2) increases
   (3) remains the same   5 _____

6. As the pressure on the surface of a liquid decreases, the temperature at which the liquid will boil

   (1) decreases
   (2) increases
   (3) remains the same   6 _____

7. Using your knowledge of chemistry and the information in Reference Table H, which statement concerning propanone and water at 50°C is true?

   (1) Propanone has a higher vapor pressure and stronger intermolecular forces than water.
   (2) Propanone has a higher vapor pressure and weaker intermolecular forces than water.
   (3) Propanone has a lower vapor pressure and stronger intermolecular forces than water.
   (4) Propanone has a lower vapor pressure and weaker intermolecular forces than water.

                                                   7 _____

8. A liquid boils when the vapor pressure of the liquid equals the atmospheric pressure on the surface of the liquid. Using Reference Table H, determine the boiling point of water when the atmospheric pressure is 90. kPa.

Base your answers to question 9 using your knowledge of chemistry and on the graph below, which shows the vapor pressure curves for liquids *A* and *B*. **Note:** The pressure is given in mm Hg – millimeters of mercury.

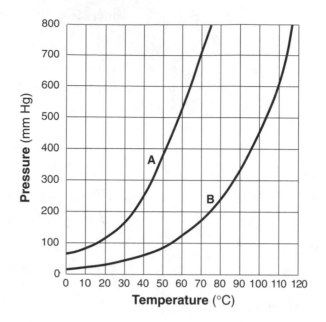

9. *a)* What is the vapor pressure of liquid *A* at 70°C? Your answer must include correct units.

_____

b) At what temperature does liquid *B* have the same vapor pressure as liquid *A* at 70°C? Your answer must include correct units.

_____

c) At 400 mm Hg, which liquid would reach its boiling point first? _____

d) Which liquid will evaporate more rapidly? Explain your answer in terms of intermolecular forces.

_____

_____

_____

Vapor Pressure of Four Liquids

10. At 65°C, which compound has a vapor pressure of 58 kilopascals?

    (1) ethanoic acid   (3) propanone
    (2) ethanol   (4) water   10 ____

11. Which liquid has the highest vapor pressure at 75°C?

    (1) ethanoic acid   (3) propanone
    (2) ethanol   (4) water   11 ____

12. When the vapor pressure of water is 70 kPa the temperature of the water is

    (1) 20°C   (3) 60°C
    (2) 40°C   (4) 91°C   12 ____

13. According to Reference Table H, what is the boiling point of ethanoic acid at 80 kPa?

    (1) 28°C   (3) 111°C
    (2) 100°C   (4) 125°C   13 ____

14. The graph below shows the relationship between vapor pressure and temperature for substance **X**.

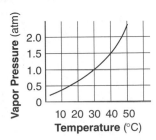

    What is the normal boiling point for substance **X**?

    (1) 50°C   (3) 30°C
    (2) 20°C   (4) 40°C   14 ____

15. The table below shows the normal boiling point of four compounds.

Compound	Normal Boiling Point (°C)
HF(ℓ)	19.4
CH$_3$Cl(ℓ)	−24.2
CH$_3$F(ℓ)	−78.6
HCl(ℓ)	−83.7

    Which compound has the strongest intermolecular forces?

    (1) HF(ℓ)   (3) CH$_3$F(ℓ)
    (2) CH$_3$Cl(ℓ)   (4) HCl(ℓ)   15 ____

16. Based on Reference Table H, which subtance has the weakest intermolecular forces?

    (1) ethanoic acid
    (2) ethanol
    (3) propanone
    (4) water   16 ____

17. The graph below represents the vapor pressure curves of four liquids.

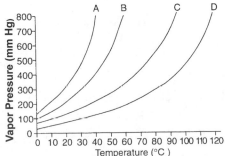

    Which liquid has the highest boiling point at 600 mm Hg?

    (1) A   (3) C
    (2) B   (4) D   17 ____

18. A liquid's boiling point is the temperature at which its vapor pressure is equal to the atmospheric pressure. Using Reference Table H, what is the boiling point of propanone at an atmospheric pressure of 70 kPa?

_45°C_

19. Explain, in terms of molecular energy, why the vapor pressure of propanone increases when its temperature increases.

As temp increases, molecules gain more energy, allowing liquid to vapor. Vapor pressure increases.

20. The boiling point of a liquid is the temperature at which the vapor pressure of the liquid is equal to the pressure on the surface of the liquid. The heat of vaporization of ethanol is 838 joules per gram. A sample of ethanol has a mass of 65.0 grams and is boiling at 1.00 atmosphere.

Based on Table H, what is the temperature of this sample of ethanol? _80°C_

21. A sample of ethanoic acid is at 85°C. At a pressure of 50 kPa, what increase in temperature is needed to reach the boiling point of ethanoic acid? _11°C_

22. Based on Reference Table H, which substance has the:

strongest intermolecular forces – _ethanoic acid_

weakest intermolecular forces – _propanone_

23. At 70 kPa, determine the boiling point of:

propanone – _____ °C

ethanol – _____ °C

water – _____ °C

ethanoic acid – _____ °C

Vapor Pressure of Four Liquids

1. 2   On Table H, locate the 75°C line. As one moves upward the first line intersected is ethanoic acid. This would have the lowest vapor pressure at this temperature, around 23 kPa.

2. 3   On Table H, go to the 45°C temperature line. Follow this line up until it intersects the propanone vapor pressure curve. The vapor pressure is 70 kPa.

3. 3   The vapor pressure line of 48 kPa intersects the propanone vapor pressure curve at a temperature of 35°C. This intersection point represents the boiling point of propanone at this pressure.

4. 2   At 35°C, propanone has a vapor pressure of 48 kPa. At 60°C, ethanol would also have a vapor pressure of 48 kPa.

5. 2   Above the surface of a liquid, some vapor will always be found. As the temperature of the liquid increases, more of that liquid will turn to vapor. At the higher temperature, the vapor particles will possess more kinetic energy and therefore cause an increase in pressure of the vapor.

6. 1   A liquid boils when the vapor pressure of the liquid equals the atmospheric pressure on the surface of the liquid. As the temperature of the liquid increases, the vapor pressure increases. If the pressure on the surface of a liquid decreases, the temperature of the liquid at which the vapor pressure will equal the atmospheric pressure will be lower. Therefore, the liquid will boil at a lower or decreased temperature.

7. 2   Open to Table H and notice that propanone has a higher vapor pressure (84 kPa) than water (12 kPa) at 50°C. The higher vapor pressure of propanone at this temperature indicates that the intermolecular forces between its molecules are weaker than that of water, allowing the molecules to escape more readily to the vapor phase.

8.   Answer: 95°C   *or*   96°C   *or*   97°C

   Explanation: Locate the 90 kPa line on the Vapor Pressure graph. Move directly over to the right until it intersects the water vapor pressure curve. Reading down to the temperature line gives its boiling point temperature at this pressure.

9. *a)* Answer: 700 mm

    Explanation: Find the 70°C value on the temperature axis of the graph and read directly up until the line intersects the vapor pressure curve of liquid *A*. Now go directly to the left to the vapor pressure axis and read the vapor pressure of liquid *A* at 70°C, which is 700 mm.

b) Answer: 114°C ± 2°C

    Explanation: On the pressure axis, find the 700 mm value. Read directly to the right until this line intersects the vapor pressure curve of liquid *B*. Now read directly down to the temperature axis. The temperature value is 114°C.

c) Answer: liquid *A*

    Explanation: Locate the 400 mm pressure value line. Move directly to the right from this value until it intersects the first line (liquid *A*). At this intersection point, liquid *A* is at its boiling point for this pressure.

d) Answer: liquid *A*

    Explanation: At any given temperature, the vapor pressure of liquid *A* is greater than that of liquid *B*. This indicates that there are more vapor particles of liquid *A* above its surface than vapor particles of liquid *B* above its surface. One can then conclude that liquid *A* evaporates more readily than liquid *B* and therefore the intermolecular forces between molecules of *A* are less than those between molecules of *B*.

       Vapor Pressure of Four Liquids

| Table I | Heats of Reaction |

Heats of Reaction at 101.3 kPa and 298 K

Reaction	ΔH (kJ)*
$CH_4(g) + 2O_2(g) \longrightarrow CO_2(g) + 2H_2O(\ell)$	−890.4
$C_3H_8(g) + 5O_2(g) \longrightarrow 3CO_2(g) + 4H_2O(\ell)$	−2219.2
$2C_8H_{18}(\ell) + 25O_2(g) \longrightarrow 16CO_2(g) + 18H_2O(\ell)$	−10943
$2CH_3OH(\ell) + 3O_2(g) \longrightarrow 2CO_2(g) + 4H_2O(\ell)$	−1452
$C_2H_5OH(\ell) + 3O_2(g) \longrightarrow 2CO_2(g) + 3H_2O(\ell)$	−1367
$C_6H_{12}O_6(s) + 6O_2(g) \longrightarrow 6CO_2(g) + 6H_2O(\ell)$	−2804
$2CO(g) + O_2(g) \longrightarrow 2CO_2(g)$	−566.0
$C(s) + O_2(g) \longrightarrow CO_2(g)$	−393.5
$4Al(s) + 3O_2(g) \longrightarrow 2Al_2O_3(s)$	−3351
$N_2(g) + O_2(g) \longrightarrow 2NO(g)$	+182.6
$N_2(g) + 2O_2(g) \longrightarrow 2NO_2(g)$	+66.4
$2H_2(g) + O_2(g) \longrightarrow 2H_2O(g)$	−483.6
$2H_2(g) + O_2(g) \longrightarrow 2H_2O(\ell)$	−571.6
$N_2(g) + 3H_2(g) \longrightarrow 2NH_3(g)$	−91.8
$2C(s) + 3H_2(g) \longrightarrow C_2H_6(g)$	−84.0
$2C(s) + 2H_2(g) \longrightarrow C_2H_4(g)$	+52.4
$2C(s) + H_2(g) \longrightarrow C_2H_2(g)$	+227.4
$H_2(g) + I_2(g) \longrightarrow 2III(g)$	+53.0
$KNO_3(s) \xrightarrow{H_2O} K^+(aq) + NO_3^-(aq)$	+34.89
$NaOH(s) \xrightarrow{H_2O} Na^+(aq) + OH^-(aq)$	−44.51
$NH_4Cl(s) \xrightarrow{H_2O} NH_4^+(aq) + Cl^-(aq)$	+14.78
$NH_4NO_3(s) \xrightarrow{H_2O} NH_4^+(aq) + NO_3^-(aq)$	+25.69
$NaCl(s) \xrightarrow{H_2O} Na^+(aq) + Cl^-(aq)$	+3.88
$LiBr(s) \xrightarrow{H_2O} Li^+(aq) + Br^-(aq)$	−48.83
$H^+(aq) + OH^-(aq) \longrightarrow H_2O(\ell)$	−55.8

*The ΔH values are based on molar quantities represented in the equations. A minus sign indicates an exothermic reaction.

Overview:

Any reaction, chemical or physical, involves either the absorption or release of energy. This energy is usually measured in the form of heat, expressed in kJ, and is called the enthalpy or simply the heat of reaction. The symbol for heat of reaction is ΔH. The conditions of 101.3 kPa and 298 K are the standard conditions for the measurement of heats of reaction.

The Table:

The chemical equations for many reactions are given on this table. Notice that the phases of the reactants and products are given in each reaction. The heats of reaction (ΔH) are given at the right hand side of the table. As indicated at the bottom of the table by the asterisk, a minus sign indicates an exothermic reaction. An exothermic reaction is one that produces or releases energy. This is indicated by a negative heat of reaction. When included in the chemical equation, the heat is included on the product side.

For example: The first reaction on the chart is an exothermic reaction since it has a negative heat of reaction (–890.4 kJ). Expressed as a chemical equation, it would be written as:

$$CH_4(g) + 2O_2(g) \rightarrow CO_2(g) + 2H_2O(\ell) + 890.4 \text{ kJ}$$

An endothermic reaction is one that absorbs energy. This is indicated by a positive heat of reaction. When included in the chemical equation, the heat is included on the reactant side.

For example: The first endothermic reaction given on the chart is $N_2(g) + O_2(g) \rightarrow 2NO(g)$ as shown by the positive heat of reaction (182.6 kJ). This reaction would absorb 182.6 kJ. Expressed as a chemical equation, it would be written as:

$$N_2(g) + O_2(g) + 182.6 \text{ kJ} \rightarrow 2NO(g)$$

Potential Energy Diagrams: Chemical reactions can be shown by a potential energy diagram. Below is a potential energy diagram of an endothermic reaction.

Endothermic Reaction

In endothermic reactions, the potential energy of the products is greater than the potential energy of the reactants. Endothermic reactions absorb heat from the surroundings.

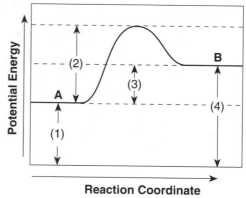

$$2\,C(s) + H_2(g) + 227.4 \text{ kJ} \rightarrow C_2H_2(g)$$

Explanation of the Arrows:

Arrow (1) represents the potential energy of the reactants, $2C(s) + H_2(g)$. This potential energy is illustrated by plateau A.

Arrow (2) is the activation energy. This is the minimum amount of energy needed to start the reaction.

Arrow (3) is the heat of reaction (ΔH). The heat of reaction is the potential energy of the product minus the potential energy of the reactant. Since B is greater than A, the ΔH for an endothermic reaction is positive ($\Delta H = 227.4$ kJ).

Notice that, in the chemical equation, the heat is included on the reactant side of the equation since it is absorbed. This is true for all endothermic reactions.

Arrow (4) represents the potential energy of the product, $C_2H_2(g)$, plateau B.

Exothermic Reaction

In exothermic reactions, the potential energy of the products is lower than the potential energy of the reactants. Exothermic reactions release heat into the surroundings.

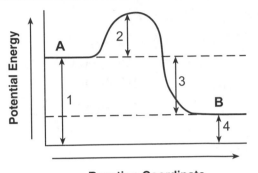

Reaction Coordinate

$$CH_4(g) + 2O_2(g) \rightarrow CO_2(g) + 2H_2O(\ell) + 890.4 \text{ kJ}$$

Arrow (1) represents the potential energy of the reactants, $CH_4(g) + 2O_2(g)$.

Arrow (2) represents the activation energy.

Arrow (3) is the heat of reaction (ΔH). Since 4 is less than 1, the ΔH for an exothermic reaction is negative ($\Delta H = -890.4$ kJ). Notice that, in the chemical equation, the heat is included on the product side of the equation since it is released. This is true for all exothermic reactions.

Arrow (4) represents the potential energy of the products, $CO_2(g) + 2H_2O(\ell)$, plateau B.

Catalyst

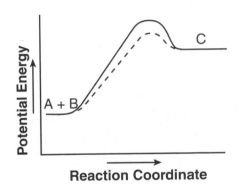

Reaction Coordinate

A catalyst is a substance that increases the speed at which a reaction takes place or equilibrium is reached. The catalyst accomplishes this by lowering the activation energy needed for the forward and reverse reaction. In the diagram to the left, the solid line represents the change in potential energy that occurs during the given reaction: $A + B \rightarrow C$. The dotted line represents the pathway of the same reaction when a catalyst is added. Notice that a different reaction pathway occurs with the peak of the potential energy graph (the top of the curve) being lowered. Thus the activation energy is decreased, without changing the positions of the potential energy of the plateaus (reactants and products).

Additional Information:

- Melting and evaporation are endothermic phase changes.
  Condensation and freezing are exothermic phase changes.

- The greater the negative ΔH, the more stable the products of the reaction.

- The greater the positive ΔH, the more unstable the products of the reaction.

- Entropy is a measure of the randomness or disorder in a system. As a system becomes more random (less ordered), the entropy increases. As the temperature of a system increases, the entropy increases. As a substance changes from a solid to a liquid to a gas, the entropy increases.

1. Which statement best describes a chemical reaction in which energy is released?

   (1) It is exothermic and has a negative ΔH.
   (2) It is exothermic and has a positive ΔH.
   (3) It is endothermic and has a negative ΔH.
   (4) It is endothermic and has a positive ΔH.

   1 _____

2. Which change of phase is exothermic?

   (1) solid to liquid (3) solid to gas
   (2) gas to liquid   (4) liquid to gas 2 _____

3. Which equation represents an exothermic reaction at 298 K?

   (1) $N_2(g) + O_2(g) \rightarrow 2NO(g)$
   (2) $C(s) + O_2(g) \rightarrow CO_2(g)$
   (3) $KNO_3(s) \rightarrow K^+(aq) + NO_3^-(aq)$
   (4) $NH_4Cl(s) \rightarrow NH_4^+(aq) + Cl^-(aq)$ 3 _____

4. Based on Reference Table I, which change occurs when pellets of solid NaOH are added to water and stirred?

   (1) The water temperature increases as chemical energy is converted to heat energy.
   (2) The water temperature increases as heat energy is stored as chemical energy.
   (3) The water temperature decreases as chemical energy is converted to heat energy.
   (4) The water temperature decreases as heat energy is stored as chemical energy. 4 _____

5. Which reaction releases the greatest amount of energy per 2 moles of product?

   (1) $2CO(g) + O_2(g) \rightarrow 2CO_2(g)$
   (2) $4Al(s) + 3O_2(g) \rightarrow 2Al_2O_3(s)$
   (3) $2H_2(g) + O_2(g) \rightarrow 2H_2O(g)$
   (4) $N_2(g) + 3H_2(g) \rightarrow 2NH_3(g)$ 5 _____

6. In a potential energy diagram, the difference between the potential energy of the products and the potential energy of the reactants is equal to the

   (1) heat of reaction
   (2) entropy of the reaction
   (3) activation energy of the forward reaction
   (4) activation energy of the reverse reaction

   6 _____

7. Which balanced equation represents an endothermic reaction?

   (1) $C(s) + O_2(g) \rightarrow CO_2(g)$
   (2) $CH_4(g) + 2O_2(g) \rightarrow CO_2(g) + 2H_2O(\ell)$
   (3) $N_2(g) + 3H_2(g) \rightarrow 2NH_3(g)$
   (4) $N_2(g) + O_2(g) \rightarrow 2NO(g)$ 7 _____

8. Which statement correctly describes an endothermic chemical reaction?

   (1) The products have higher potential energy than the reactants, and the ΔH is negative.
   (2) The products have higher potential energy than the reactants, and the ΔH is positive.
   (3) The products have lower potential energy than the reactants, and the ΔH is negative.
   (4) The products have lower potential energy than the reactants, and the ΔH is positive. 8 _____

9. According to Table I, which salt releases energy as it dissolves?

(1) KNO_3 (3) NH_4NO_3
(2) LiBr   (4) NaCl   9 _____

10. Given the balanced equation:

$$KNO_3(s) + 34.89 \text{ kJ} \xrightarrow{H_2O} K^+(aq) + NO_3^-(aq)$$

Which statement best describes this process?

(1) It is endothermic and entropy increases.
(2) It is endothermic and entropy decreases.
(3) It is exothermic and entropy increases.
(4) It is exothermic and entropy decreases.

10 _____

11. Given the potential energy diagram of a chemical reaction:

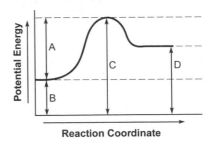

Which arrow represents the potential energy of the reactants?

(1) A   (3) C
(2) B   (4) D   11 _____

12. Given the potential energy diagram for a chemical reaction. Which statement correctly describes the energy changes that occur in the forward reaction?

(1) The activation energy is 10. kJ and the reaction is endothermic.
(2) The activation energy is 10. kJ and the reaction is exothermic.
(3) The activation energy is 50. kJ and the reaction is endothermic.
(4) The activation energy is 50. kJ and the reaction is exothermic.

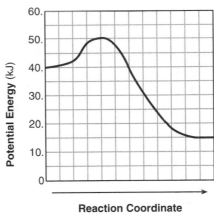

12 _____

13. Given the balanced equation representing a reaction:

$$H_2(g) + Cl_2(g) \rightarrow 2HCl(g) + energy$$

Which statement describes the energy changes in this reaction?

(1) Energy is absorbed as bonds are formed, only.
(2) Energy is released as bonds are broken, only.
(3) Energy is absorbed as bonds are broken, and energy is released as bonds are formed.
(4) Energy is absorbed as bonds are formed, and energy is released as bonds are broken.   13 _____

Base your answer to question 14 using the information below and your knowledge of chemistry.

Given the balanced equation for dissolving $NH_4Cl(s)$ in water:

$$NH_4Cl(s) \xrightarrow{H_2O} NH_4^+(aq) + Cl^-(aq)$$

14. A student is holding a test tube containing 5.0 milliliters of water. A sample of $NH_4Cl(s)$ is placed in the test tube and stirred. Describe the heat flow between the test tube and the student's hand.

_____

_____

Base your answers to question 15 on the information below.

The catalytic converter in an automobile changes harmful gases produced during fuel combustion to less harmful exhaust gases. In the catalytic converter, nitrogen dioxide reacts with carbon monoxide to produce nitrogen and carbon dioxide. This reaction is represented by the balanced equation below.

Reaction 1: $2NO_2(g) + 4CO(g) \rightarrow N_2(g) + 4CO_2(g) + 1198.4$ kJ

15. The accompanying potential energy diagram represents reaction 1 without a catalyst. On the same diagram, draw a dashed line to indicate how potential energy changes when the reaction is catalyzed in the converter.

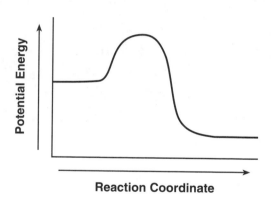

Given the balanced equation representing a reaction:

$$N_2(g) + O_2(g) + 182.6 \text{ kJ} \rightarrow 2NO(g)$$

16. a) On the accompanying labeled axes, draw a potential energy diagram for this reaction.

b) What is the ΔH for this reaction?

_____ kJ

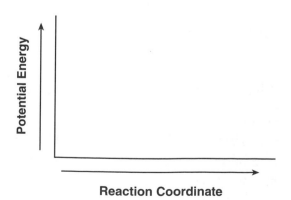

17. Which phase change is endothermic?

    (1) gas → solid   (3) liquid → solid
    (2) gas → liquid   (4) liquid → gas   17 ____

18. Of the following reactions, which one releases the most heat?

    (1) $CH_4(g) + 2O_2(g) \rightarrow CO_2(g) + 2H_2O(\ell)$
    (2) $2C_8H_{18}(\ell) + 25O_2(g) \rightarrow 16CO_2(g) + 18H_2O(\ell)$
    (3) $N_2(g) + O_2(g) \rightarrow 2NO(g)$
    (4) $2C(s) + H_2(g) \rightarrow C_2H_2(g)$ 18 ____

19. Given the balanced equation representing a reaction:

    $CH_4(g) + 2O_2(g) \rightarrow 2H_2O(\ell) + CO_2(g) + heat$

    Which statement is true about energy in this reaction?

    (1) The reaction is exothermic because it releases heat.
    (2) The reaction is exothermic because it absorbs heat.
    (3) The reaction is endothermic because it releases heat.
    (4) The reaction is endothermic because it absorbs heat.   19 ____

20. Given the balanced equation representing a reaction at 101.3 kPa and 298 K:

    $N_2(g) + 3H_2(g) \rightarrow 2NH_3(g) + 91.8 kJ$

    Which statement is true about this reaction?

    (1) It is exothermic and ΔH equals –91.8 kJ.
    (2) It is exothermic and ΔH equals +91.8 kJ.
    (3) It is endothermic and ΔH equals –91.8 kJ.
    (4) It is endothermic and ΔH equals +91.8 kJ.   20 ____

21. When lithium bromide crystals are dissolved in water, the temperature of the water increases. What does this temperature change indicate about the dissolving of lithium bromide in water?

    (1) It is an endothermic reaction because it absorbs heat.
    (2) It is an endothermic reaction because it releases heat.
    (3) It is an exothermic reaction because it absorbs heat.
    (4) It is an exothermic reaction because it releases heat.   21 ____

22. Given the reaction:

    $2H_2(g) + O_2(g) \rightarrow 2H_2O(\ell) + 571.6 kJ$

    What is the approximate ΔH for the formation of 1 mole of $H_2O(\ell)$?

    (1) –285.8 kJ   (3) –571.6 kJ
    (2) +285.8 kJ   (4) +571.6 kJ   22 ____

23. Given the potential energy diagram and equation representing the reaction between substances A and D:

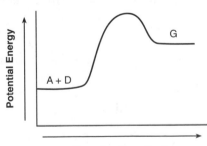

Reaction Coordinate

$A + D \longrightarrow G$

According to Table I, substance G could be

    (1) HI(g)   (3) $CO_2(g)$
    (2) $H_2O(g)$ (4) $C_2H_6(g)$ 23 ____

24. For a given reaction, adding a catalyst increases the rate of the reaction by
    (1) providing an alternate reaction pathway that has a higher activation energy
    (2) providing an alternate reaction pathway that has a lower activation energy
    (3) using the same reaction pathway and increasing the activation energy
    (4) using the same reaction pathway and decreasing the activation energy 24____

25. Which phase change results in the release of energy?
    (1) $H_2O(s) \rightarrow H_2O(\ell)$
    (2) $H_2O(s) \rightarrow H_2O(g)$
    (3) $H_2O(\ell) \rightarrow H_2O(g)$
    (4) $H_2O(g) \rightarrow H_2O(\ell)$ 25 ____

26. Given the balanced equation representing a phase change:

    $$C_6H_4Cl_2(s) + energy \rightarrow C_6H_4Cl_2(g)$$

    Which statement describes this change?
    (1) It is endothermic, and entropy decreases.
    (2) It is endothermic, and entropy increases.
    (3) It is exothermic, and entropy decreases.
    (4) It is exothermic, and entropy increases.
                                          26 ____

27. Which balanced equation represents a chemical change?
    (1) $H_2O(\ell) + energy \rightarrow H_2O(g)$
    (2) $2H_2O(\ell) + energy \rightarrow 2H_2(g) + O_2(g)$
    (3) $H_2O(\ell) \rightarrow H_2O(s) + energy$
    (4) $H_2O(g) \rightarrow H_2O(\ell) + energy$ 27 ____

Base your answers to question 28 using the information below and your knowledge of chemistry.

Propane is a fuel that is sold in rigid, pressurized cylinders. Most of the propane in a cylinder is liquid, with gas in the space above the liquid level. When propane is released from the cylinder, the propane leaves the cylinder as a gas. Propane gas is used as a fuel by mixing it with oxygen in the air and igniting the mixture, as represented by the balanced equation below.

$$C_3H_8(g) + 5O_2(g) \rightarrow 3CO_2(g) + 4H_2O(\ell) + 2219.2 \text{ kJ}$$

28. a) Determine the total amount of energy released when
    2.50 moles of propane is completely reacted with oxygen. _____ kJ

    b) Is the above reaction exothermic or endothermic? _____

Base your answers to question 29 on the information below and your knowledge of chemistry.

A hot pack contains chemicals that can be activated to produce heat. A cold pack contains chemicals that feel cold when activated.

29. a) Based on energy flow, state the type of chemical change that occurs in a hot pack.

_____

    b) A cold pack is placed on an injured leg. Indicate the direction of the flow of energy between the leg and the cold pack.

_____

Heats of Reaction   Copyright © 2012
Topical Review Book Company

Base your answer to question 30 using your knowledge of chemistry and the information and accompanying diagram, which represent the changes in potential energy that occur during the given reaction.

Given the reaction: $A + B \rightarrow C$

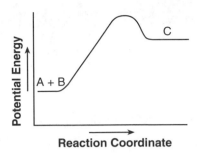

30. Does the diagram illustrate an exothermic or an endothermic reaction? State one reason, in terms of energy, to support your answer.

_____

_____

Base your answers to question 31 on the potential energy diagram.

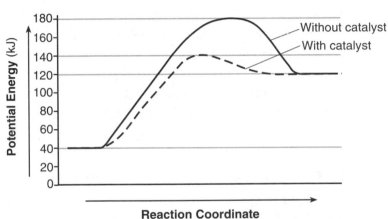

31. *a*) What is the heat of reaction for the forward reaction?

_____ kJ

b) What is the activation energy for the forward reaction with the catalyst?

_____ kJ

c) Explain, in terms of the function of a catalyst, why the curves on the potential energy diagram for the catalyzed and uncatalyzed reactions are different.

_____

_____

32. Explain what will happen to the thermometer reading when 22 grams of NaOH(s) is dissolved in the water.

_____

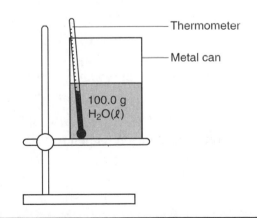

Table I – Heats of Reaction
Answers
Set 1

1. 1   All exothermic reactions release heat and have a negative (minus) ΔH. This information is given at the bottom of Table I.

2. 2   Melting and evaporation are phase changes that absorb heat making them endothermic phase changes. Condensation and freezing are phase changes that release heat making them exothermic phase changes. Condensation is when a gas changes to a liquid.

3. 2   An exothermic reaction releases energy. Open to Table I. It shows that $C(s) + O_2(g) \rightarrow CO_2(g)$ releases 393.5 kJ of heat, indicated by a negative ΔH. At the bottom of this chart it states "Minus sign indicates an exothermic reaction".

4. 1   In Table I, locate the equation involving the reactant NaOH. The ΔH for this reaction is -44.51 kJ, making this an exothermic reaction. This reaction releases heat as chemical energy, which converted to heat energy.

5. 2   When a reaction releases energy it is an exothermic reaction, having a negative ΔH. The reaction shown in choice 2 releases 3351 kJ of energy per 2 moles of product. This is the greatest amount of energy of the given choices.

6. 1   The reactants involved in a chemical reaction will be at a specific potential energy, measured in kilojoules (kJ). After the reaction, the products will be at a different potential energy level. The net difference between these two energy levels, as shown by two plateaus on a potential energy graph, is the referred to as the heat of reaction.

7. 4   In Table I, locate this equation. Its ΔH is $+182.6$ kJ. All endothermic reactions have a positive ΔH.

8. 2   When the products have a higher potential energy than the reactants, the chemical reaction must have absorbed energy, which makes it an endothermic reaction. A graph would show the reactants at a lower potential energy than the products.

9. 2   As shown on Table I, when LiBr(s) is dissolved, an exothermic reaction takes place releasing 48.83 kJ of energy. All other choices are endothermic reactions, which would absorb heat.

10. 1 The equation shows that 34.89 kJ of energy being absorbed by the reactants, making it endothermic reaction. Entropy is a measure of the disorder of a system. The greater the disorder, the greater the entropy. The equation shows that the solid KNO_3 is being dissolved to form an aqueous solution. Whenever a solid is dissolved, entropy increases.

11. 2   The potential energy of the reactants is shown by the initial plateau, which is letter B. The potential energy of the products is shown by the second plateau, letter D.

12. 2   The graph shows that it took 10 kJ (40 kJ to 50 kJ) of activation energy for this reaction to occur. The graph also shows that the potential energy of the products is lower then that of the reactants. This is true for all exothermic reactions.

13. 3   The equation shows the reaction between two diatomic molecules (H_2 and Cl_2) to produce a single product (HCl). The bonds between the Cl atoms in Cl_2 and the H atoms in H_2 must first be broken, which requires the absorption of energy. The separate H atoms and Cl atoms then bond to each other to form the product, HCl. Bond formation releases energy.

14.   Answer: Heat flows from the student's hand to the test tube.
        or   The test tube absorbs heat from the hand.

        Explanation: As shown in Table I, this reaction is endothermic absorbing 14.78 kJ. The reaction will absorb heat from the water and from the student's hand. As heat is removed (being absorbed) from the student's hand, the test tube will feel colder.

15.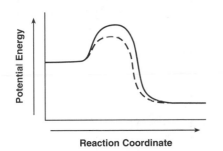

        Reaction Coordinate

        Explanation: Refer to page 41 and the given explanation for the catalyst diagram.

16. *a)*

        Reaction Coordinate

        Explanation: The given equation is an endothermic reaction since the heat is included on the reactant side of the chemical equation. On a graph, the potential energy of the products must be greater than the potential energy of the reactants for endothermic reactions.

    b)   Answer: + 182.6 kJ

        Explanations: This is an endothermic reaction. Endothermic reactions will always have a positive heat of reaction (ΔH).

Overview:

You have probably noticed that if iron is not painted or coated, it starts to rust in days, especially if moisture is present. But silver, and especially gold, seem to be unaffected by substances in the environment and keep their brilliant luster. The reason for this is that different metals exhibit different chemical activity. In other words, some metals are very reactive, while other metals are less reactive. By studying the chemical activities of elements (metals and nonmetals), chemists have been able to arrange them based upon chemical reactivity.

The Table:

This table shows the relative chemical activity of metals and nonmetals, both arranged in order of decreasing chemical activity. Although H_2 is not a metal, it is listed on the metallic side because the table is based on the hydrogen standard.

In a chemical reaction, a more active metal (higher up on Table J) will replace a less active metal when placed in an aqueous solution containing the ion of the less active metal. For example, lithium (Li), being the most active metal, will replace any metallic ion found below it from a solution of its salt. Rubidium (Rb) will replace any metal found below it from a solution of its salt, but because it is under Li, indicating that it is less active than Li, it will not replace Li from a solution of its salt.

Metals found above H_2 are more active than hydrogen. Therefore, it will replace the H^+ in an aqueous acidic solution, producing hydrogen gas and a solution of a salt containing that metal. Those metals below H_2 will not react with acids in this fashion.

Most Active	Metals	Nonmetals	Most Active
	Li	F_2	
	Rb	Cl_2	
	K	Br_2	
	Cs	I_2	
	Ba		
	Sr		
	Ca		
	Na		
	Mg		
	Al		
	Ti		
	Mn		
	Zn		
	Cr		
	Fe		
	Co		
	Ni		
	Sn		
	Pb		
	H_2		
	Cu		
	Ag		
Least Active	Au		Least Active

**Activity Series is based on the hydrogen standard. H_2 is *not* a metal.

For example: $Mg + 2HCl \rightarrow MgCl_2 + H_2\uparrow$ reacted, because Mg is more active (being above H_2) than H_2, as shown on Table J.

$Ag + HCl \rightarrow$ no reaction, because Ag is less active (being lower than H_2) than H_2, as shown on Table J.

In a similar fashion, in a chemical reaction, a nonmetal will replace a less active nonmetal when interacting with a solution containing the ion of the less active nonmetal. For example, fluorine (F_2) will replace Cl^-, Br^- and I^- in solutions containing those ions, but chlorine (Cl_2) cannot replace F^- from a solution containing the F^- ion; however, chlorine will replace Br^- and I^- in solutions containing these ions.

Voltaic Cell:

An electrochemical or voltaic cell uses a spontaneous redox reaction to produce an electric current. It consists of two different metals, called electrodes, immersed in a solution of that metals salt, called an electrolyte. The electrodes are connected by a wire conductor. The electrolytes are connected by a salt bridge. The more active metal, higher up on Table J, undergoes oxidation and is called the anode (the negative electrode). The electrons flow through the wire to the less active metal, lower down on Table J, where they reduce that metals ions in the electrolyte. This electrode is called the cathode (the positive electrode).

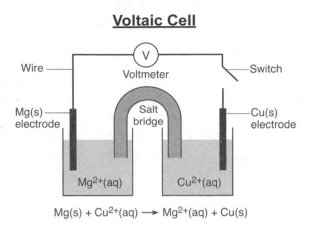

Voltaic Cell

$$Mg(s) + Cu^{2+}(aq) \longrightarrow Mg^{2+}(aq) + Cu(s)$$

Example:

A voltaic cell with magnesium and copper electrodes is shown in the above diagram. The copper electrode has a mass of 15.0 grams. Below the diagram is the balanced ionic equation for the reaction in the cell.

When the switch is closed, the salt bridge allows ions to flow between the half-cells and the reaction in the cell begins. The more active metal, Mg being higher up on Table J, will undergo oxidation (acting as a reducing agent) and the less active metal, Cu will undergo reduction (acting as an oxidizing agent). The electrons flow from the Mg electrode (losing electrons during oxidation) to the copper electrode where they will be used in the reduction of the Cu^{2+} ions found in the electrolyte. As Cu^{2+} ions become reduced to Cu^o (a neutral copper atom), the atoms become part of the copper electrode, increasing its mass.

In an electrolytic cell, an electric current is used to cause a nonspontaneous redox reaction to occur. It needs a power source, such as a battery, to begin and sustain the reaction. In this reaction, electrical energy is converted to chemical energy. In the voltaic cell, chemical energy is converted to electrical energy.

Review:

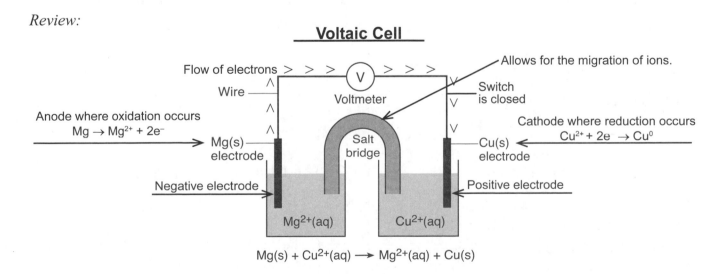

Voltaic Cell

$$Mg(s) + Cu^{2+}(aq) \longrightarrow Mg^{2+}(aq) + Cu(s)$$

Additional Information:

- In the single replacement reaction $A + BC \rightarrow B + AC$ where A is a metal, the reaction will occur spontaneously if A is above B on Table J. If A is below B, the reaction will not occur.

- The most active metals are those that readily lose an electron, thus are easily oxidized. They tend to be the strongest reducing agents. Typically, they are Group 1 and Group 2 elements.

- In the single replacement reaction $A + BC \rightarrow C + BA$ where A is a nonmetal, the reaction will occur spontaneously if A is above C on Table J. If A is below C, the reaction will not occur.

- The most active nonmetals are those that more readily gain an electron, thus most easily reduced. They tend to be the strongest oxidizing agents. Typically, they are Group 17 elements.

- Fluorine is the most active of all the elements.

- Gold (Au) a very inactive metal, which is why it keeps its brilliant luster. Being inactive, it is used in electrical connections that are exposed to hostile conditions, such as those found in space vehicles.

Set 1 — Activity Series

1. According to Reference Table J, which of these metals will react most readily with 1.0 M HCl to produce $H_2(g)$?

   (1) Ca   (3) Mg
   (2) K   (4) Zn   1 _____

2. Which metal is more active than H_2?

   (1) Ag   (3) Cu
   (2) Au   (4) Pb   2 _____

3. According to Reference Table J, which metal will react with Zn^{2+} but will not react with Mg^{2+}?

   (1) Al(s)   (3) Ni(s)
   (2) Cu(s)   (4) Ba(s)   3 _____

4. Which metal reacts spontaneously with a solution containing zinc ions?

   (1) strontium   (3) copper
   (2) nickel   (4) silver   4 _____

5. Which of the following metals is most active?

   (1) Ag   (3) Sn
   (2) Zn   (4) Li   5 _____

6. Which metal is more active than Ni and less active than Zn?

   (1) Cu   (3) Cr
   (2) Mg   (4) Pb   6 _____

7. Which half-reaction equation represents the reduction of an iron(II) ion?

   (1) $Fe^{2+} \rightarrow Fe^{3+} + e^-$
   (2) $Fe^{2+} + 2e^- \rightarrow Fe$
   (3) $Fe^{3+} + e^- \rightarrow Fe^{2+}$
   (4) $Fe \rightarrow Fe^{2+} + 2e^-$ 7 _____

8. The diagram below represents an operating electrochemical cell and the balanced ionic equation for the reaction occurring in the cell.

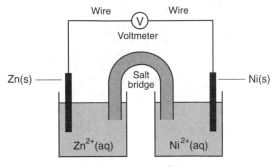

$$Zn(s) + Ni^{2+}(aq) \rightarrow Zn^{2+}(aq) + Ni(s)$$

Which statement identifies the part of the cell that conducts electrons and describes the direction of electron flow as the cell operates?

(1) Electrons flow through the salt bridge from the Ni(s) to the Zn(s).
(2) Electrons flow through the salt bridge from the Zn(s) to the Ni(s).
(3) Electrons flow through the wire from the Ni(s) to the Zn(s).
(4) Electrons flow through the wire from the Zn(s) to the Ni(s).   8 _____

9. Which statement is true for any electrochemical cell?

(1) Oxidation occurs at the anode, only.
(2) Reduction occurs at the anode, only.
(3) Oxidation occurs at both the anode and the cathode.
(4) Reduction occurs at both the anode and the cathode.   9 _____

10. Which energy conversion occurs during the operation of an electrolytic cell?

(1) chemical energy to electrical energy
(2) electrical energy to chemical energy
(3) nuclear energy to electrical energy
(4) electrical energy to nuclear energy   10 _____

11. Which process occurs at the anode in an electrochemical cell?

(1) the loss of protons
(2) the loss of electrons
(3) the gain of protons
(4) the gain of electrons   11 _____

12. Given the balanced ionic equation representing the reaction in an operating voltaic cell:

$$Zn(s) + Cu^{2+}(aq) \rightarrow Zn^{2+}(aq) + Cu(s)$$

The flow of electrons through the external circuit in this cell is from the

(1) Cu anode to the Zn cathode
(2) Cu cathode to the Zn anode
(3) Zn anode to the Cu cathode
(4) Zn cathode to the Cu anode   12 _____

13. A student collects the materials and equipment below to construct a voltaic cell.

• two 250-mL beakers
• wire and a switch
• one strip of magnesium
• one strip of copper
• 125 mL of 0.20 M $Mg(NO_3)_2(aq)$
• 125 mL of 0.20 M $Cu(NO_3)_2(aq)$

Which additional item is required for the construction of the voltaic cell?

(1) an anode
(2) a battery
(3) a cathode
(4) a salt bridge   13 _____

Activity Series

Page 53

14. Identify one metal from Reference Table J that is more easily oxidized than Mg(s). _____

Base your answer to question 15 using the information below and your knowledge of chemistry.

Two chemistry students each combine a different metal with hydrochloric acid. Student A uses zinc, and hydrogen gas is readily produced. Student B uses copper, and no hydrogen gas is produced.

15. State one chemical reason for the different results of students A and B.

_____

_____

16. When a nickel-cadmium battery produces electricity, the following reaction takes place:

$$Cd(s) + NiO_2(s) + 2H_2O(\ell) \rightarrow Cd(OH)_2(s) + Ni(OH)_2(s).$$

Explain why Cd would be above Ni if placed on Table J.

_____

_____

Base your answers to question 17 on the information below.

The diagram and balanced ionic equation represent a voltaic cell with copper and silver electrodes and the reaction that occurs when the cell is operating.

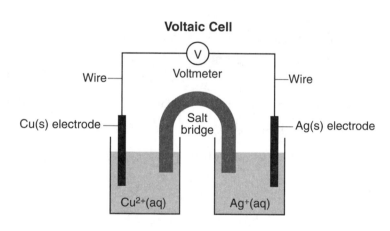

Voltaic Cell

$$Cu(s) + 2Ag^+(aq) \longrightarrow Cu^{2+}(aq) + 2Ag(s)$$

17. a) Describe the direction of electron flow in the external circuit in this operating cell.

_____

b) State the purpose of the salt bridge in this voltaic cell.

_____

c) Write a balanced half-reaction equation for the oxidation that occurs in this cell.

_____

18. Based on Reference Table J, which metal will react spontaneously with Al^{3+}?

    (1) Co(s)
    (3) Cu(s)
    (2) Cr(s)
    (4) Ca(s)

    18 _____

19. Given the balanced equation representing a reaction occurring in an electrolytic cell:

    $$2NaCl(\ell) \rightarrow 2Na(\ell) + Cl_2(g)$$

    Where is Na(ℓ) produced in the cell?
    (1) at the anode, where oxidation occurs
    (2) at the anode, where reduction occurs
    (3) at the cathode, where oxidation occurs
    (4) at the cathode, where reduction occurs

    19 _____

20. Under standard conditions, which metal will react with 0.1 M HCl to liberate hydrogen gas?

    (1) Ag
    (3) Cu
    (2) Au
    (4) Mg

    20 _____

21. Which reaction occurs spontaneously?

    (1) $Cl_2(g) + 2NaBr(aq) \rightarrow Br_2(\ell) + 2NaCl(aq)$
    (2) $Cl_2(g) + 2NaF(aq) \rightarrow F_2(g) + 2NaCl(aq)$
    (3) $I_2(s) + 2NaBr(aq) \rightarrow Br_2(\ell) + 2NaI(aq)$
    (4) $I_2(s) + 2NaF(aq) \rightarrow F_2(g) + 2NaI(aq)$

    21 _____

22. Which of the following metals has the least tendency to undergo oxidation?

    (1) Ag
    (3) Zn
    (2) Pb
    (4) Li

    22 _____

23. Which of the following nonmetals is most active?

    (1) F_2
    (3) Br_2
    (2) Cl_2
    (4) I_2

    23 _____

24. Which half-reaction correctly represents reduction?

    (1) $Mn^{4+} \rightarrow Mn^{3+} + e^-$
    (2) $Mn^{4+} \rightarrow Mn^{7+} + 3e^-$
    (3) $Mn^{4+} + e^- \rightarrow Mn^{3+}$
    (4) $Mn^{4+} + 3e^- \rightarrow Mn^{7+}$

    24 _____

25. Which energy change occurs in an operating voltaic cell?

    (1) chemical to electrical
    (2) electrical to chemical
    (3) chemical to nuclear
    (4) nuclear to chemical

    25 _____

26. Given the balanced equation representing the reaction occurring in a voltaic cell:

    $$Zn(s) + Pb^{2+}(aq) \rightarrow Zn^{2+}(aq) + Pb(s)$$

    In the completed external circuit, the electrons flow from

    (1) Pb(s) to Zn(s)
    (2) $Pb^{2+}(aq)$ to $Zn^{2+}(aq)$
    (3) Zn(s) to Pb(s)
    (4) $Zn^{2+}(aq)$ to $Pb^{2+}(aq)$

    26 _____

27. The accompanying diagram shows a key being plated with copper in an electrolytic cell. Given the reduction reaction for this cell:

$$Cu^{2+}(aq) + 2e^- \rightarrow Cu(s)$$

This reduction occurs at

(1) A, which is the anode
(2) A, which is the cathode
(3) B, which is the anode
(4) B, which is the cathode

27_____

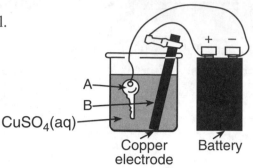

Base your answer to question 28 using the information below and your knowledge of chemistry.

The outer structure of the Statue of Liberty is made of copper metal. The framework is made of iron. Over time, a thin green layer (patina) forms on the copper surface.

28. Where the iron framework came in contact with the copper surface, a reaction occurred in which iron was oxidized. Using information from Reference Table J, explain why the iron was oxidized.

_____

_____

Base your answers to question 29 on the information below.

Underground iron pipes in contact with moist soil are likely to corrode. This corrosion can be prevented by applying the principles of electrochemistry. Connecting an iron pipe to a magnesium block with a wire creates an electrochemical cell. The magnesium block acts as the anode and the iron pipe acts as the cathode. A diagram of this system is shown to the right.

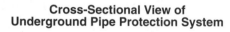

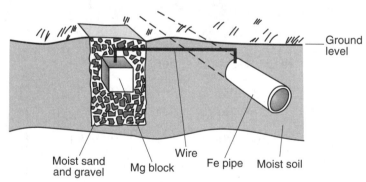

29. a) State the direction of the flow of electrons between the electrodes in this cell.

_____

b) Explain, in terms of reactivity, why magnesium is preferred over zinc to protect underground iron pipes. Your response must include both magnesium and zinc.

_____

_____

Base your answers to question 30 on the information below.

The accompanying diagram represents an operating voltaic cell at 298 K and 1.0 atmosphere in a laboratory investigation. The reaction occurring in the cell is represented by the balanced ionic equation below the diagram.

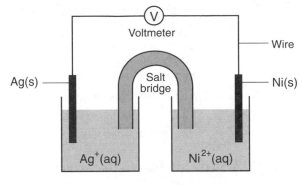

30. *a)* Identify the anode in this cell.

_____

$$2Ag^+(aq) + Ni(s) \longrightarrow 2Ag(s) + Ni^{2+}(aq)$$

b) Determine the total number of moles of $Ni^{2+}(aq)$ ions produced when 4.0 moles of $Ag^+(aq)$ ions completely react in this cell. _____

c) Write a balanced half-reaction equation for the reduction that occurs in this cell.

_____

31. Based on Table J, identify one metal that does not react spontaneously with HCl(aq). _____

32. Because tap water is slightly acidic, water pipes made of iron corrode over time, as shown by the balanced ionic equation below:

$$2Fe + 6H^+ \rightarrow 2Fe^{3+} + 3H_2$$

Explain, in terms of chemical reactivity, why copper pipes are less likely to corrode than iron pipes.

_____

33. Explain, in terms of electrical energy, how the operation of a voltaic cell differs from the operation of an electrolytic cell used in the Hall process. Include both the voltaic cell and the electrolytic cell in your answer.

_____

34. Identify *one* metal from Reference Table J that is more easily oxidized than Ba(s). _____

Activity Series

Page 57

Base your answers to question 35 on the diagram and balanced equation, which represent the electrolysis of molten NaCl.

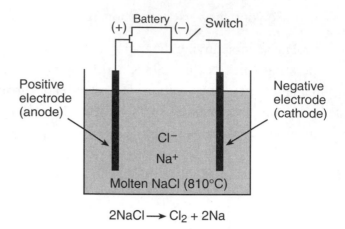

35. *a)* When the switch is closed, which electrode will attract the sodium ions?

_____

b) What is the purpose of the battery in this electrolytic cell?

_____

c) Write the balanced half-reaction for the reduction that occurs in this electrolytic cell.

_____

Base your answers to question 36 on the information below.

In a laboratory investigation, a student constructs a voltaic cell with iron and copper electrodes. Another student constructs a voltaic cell with zinc and iron electrodes. Testing the cells during operation enables the students to write the balanced ionic equations below.

Cell with iron and copper electrodes: $Cu^{2+}(aq) + Fe(s) \rightarrow Cu(s) + Fe^{2+}(aq)$

Cell with zinc and iron electrodes: $Fe^{2+}(aq) + Zn(s) \rightarrow Fe(s) + Zn^{2+}(aq)$

36. *a)* State evidence from the balanced equation for the cell with iron and copper electrodes that indicates the reaction in the cell is an oxidation-reduction reaction.

_____

b) Identify the particles transferred between Fe^{2+} and Zn during the reaction in the cell with zinc and iron electrodes.

_____

c) Write a balanced half-reaction equation for the reduction that takes place in the cell with zinc and iron electrodes.

_____

d) State the relative activity of the three metals used in these two voltaic cells.

_____

Activity Series

1. 2   The more active a metal is, the more readily it will react with a 1.0 M solution of HCl. In Table J, K (potassium) is the highest up on the metal activity list and is therefore more active than the other metals given as choices.

2. 4   Metals found above H_2 on Table J are more active than hydrogen. This would be lead (Pb).

3. 1   Any metal that is more active (higher up on Table J) will react with any metallic ion that is less active. Al is higher than Zn on the metal activity list, thus it will react with Zn^{2+}. However, Al is lower on the metal activity list than Mg, therefore it would not react with Mg^{2+}.

4. 1   A spontaneous reaction will occur with zinc ions with any metal that is higher than Zn on Table J. Strontium would have a spontaneous reaction with zinc ions.

5. 4   The most active metal of the given choices would be the one that is higher up on Table J. This would be lithium (Li).

6. 3   Using Table J, the higher up a metal is on this table, the more active it is. Notice that Cr is between Zn and Ni. This makes Cr more active than Ni, but less active than Zn.

7. 2   In a reduction reaction, electrons are gained and the oxidation state is lowered or reduced. In choice 2, Fe^{2+} gains 2 electrons, becoming Fe^o.

8. 4   Electrons flow from the electrode where oxidation or a loss of electrons occurs to the electrode where reduction or a gain of electrons occurs. The equation shows that Zn(s) is oxidized and the Ni^{2+} is reduced. The electrons therefore flow from the Zn(s) to the Ni(s), allowing the Ni^{2+} to be reduced. Ions, not electrons, flow through the salt bridge.

9. 1   In any chemical cell (voltaic or electrolytic cell), oxidation (loss of electrons) always occurs at the anode.

10. 2   See last paragraph on page 51.

11. 2   The anode is the electrode where oxidation occurs. Here, a metal gives up or loses electrons.

12. 3   Electrons flow from the anode, where oxidation occurs, to the cathode, where reduction occurs. In the equation, Zn is undergoing oxidation by losing two electrons, thus Zn is the anode. The Cu^{2+} is gaining these electrons, therefore the Cu(s) is the cathode.

13. 4   A salt bridge is an essential part of a voltaic cell. It allows the flow of ions between each half-cell, completing the circuit, allowing the cell to function.

14.   Answer: Li *or* Ba *or* Rb *or* Sr *or* K *or* Ca *or* Cs *or* Na

      Explanation: Any metal that is higher up on Table J than Mg would be more active, thus more easily oxidized than Mg.

15.   Answer: Cu is less active than hydrogen gas (weaker reducing agent).
      or Zn is more reactive (stronger reducing agent than H_2 gas).
      or Cu is below H_2 on the activity series and Zn is above H_2.

      Explanation: Any metal that is above hydrogen on Table J will react with hydrochloric acid releasing hydrogen gas. Any metal below hydrogen will not react with hydrochloric acid and will not produce hydrogen gas. Therefore, student *A*, using Zn, produced hydrogen gas and student *B*, using copper, did not have a reaction with hydrochloric acid.

16.   Answer: Cd is more active than Ni. *or* Cd oxidizes in the presence of Ni^{4+}.

      Explanation: The higher up a metal is on Table J, the more active it is and the greater its tendency to undergo oxidation. In this reaction, Cd is shown to be more active than Ni since it undergoes oxidation, replacing Ni.

17. *a)* Answer: From the copper electrode to the silver electrode.

      Explanation: Electrons flow from the anode, where oxidation occurs, to the cathode, where reduction occurs. The equation shows that copper loses electrons and therefore must be the anode.

   b) Answer: Migration of ions *or* allows ions to flow between half-cells *or* maintains electrical neutrality *or* prevents polarization

      Explanation: The salt bridge allows migration of ions between the half-cells, completing the circuit, allowing the cell to function.

   c) Answer: $Cu^0 \rightarrow Cu^{2+} + 2e^-$

      Explanation: In oxidation there is a loss of electrons. The copper metal loses two electrons, becoming a copper ion.

Formula	Name
HCl(aq)	hydrochloric acid
HNO_2(aq)	nitrous acid
HNO_3(aq)	nitric acid
H_2SO_3(aq)	sulfurous acid
H_2SO_4(aq)	sulfuric acid
H_3PO_4(aq)	phosphoric acid
H_2CO_3(aq) or CO_2(aq)	carbonic acid
CH_3COOH(aq) or $HC_2H_3O_2$(aq)	ethanoic acid (acetic acid)

Overview:

Most chemistry students are aware of the corrosiveness and danger of strong acids like sulfuric and hydrochloric acid. Acids are a class of compounds that possess or exhibit a characteristic set of physical and chemical properties. A Swedish chemist, Svante Arrhenius, investigated the behavior of acids and developed a theory of what substances would be classified as acids. The Arrhenius theory defines an acid as a substance that yields the hydrogen ion (H^+) as the only positive ion in aqueous solution. The hydrogen ion (a proton) combines with a water molecule to form the hydronium ion (H_3O^+), see Table E. The hydronium ion can also be written as H^+(aq) and is referred to a hydrated proton. This is the ion that gives all aqueous solutions of acids their characteristic properties.

The Table:

This table gives the Formula and Name of some of the more common acids one encounters in high school chemistry and everyday life. Notice that the table shows two formulas for carbonic acid. This is because carbonic acid is an aqueous solution of carbon dioxide (CO_2) gas and therefore it may be written as CO_2 (aq). Ethanoic acid, an organic acid also known as acetic acid, is the acid in vinegar. Its formula may also be written in two different ways, as shown on the table. The strength of the acid is not implied by this chart.

Additional Information:

- As a class of compounds, acids exhibit a characteristic set of properties:
  - an aqueous solution of an acid conducts an electric current (is an electrolyte)
  - acids react with certain metals (those above H_2 on Table J) to produce H_2 gas and a salt solution - Example: $Mg + 2HCl \rightarrow MgCl_2 + H_2 \uparrow$
  - acids neutralize bases (hydroxides – see Table L)
  - acids cause color changes in acid-base indicators (see Table M)
  - dilute aqueous solutions of acids have a sour taste

- Acidic solutions have a pH less than 7. The lower the pH, the stronger the acid. The stronger the acid, the greater the H^+ ion concentration. See page 74 for an explanation on the pH scale.

- In the neutralization reaction between an acid and a base, the products are a salt and water.

- The Bronsted-Lowry theory defines an acid as a substance that can donate a proton (H^+) to another substance.

- Note that the first formula given for acetic acid, CH_3COOH ends in –OH. At first glance, one might think that it is a base because of this (see Table L). Acetic acid is an organic acid and contains the functional group –COOH (see Table R). Only the H ionizes from this group, making an acidic solution.

- If the acid name ends in –ic, it is modified to end in –ate to name the ion. Sulfuric acid (H_2SO_4) gives rise to the sulfate ion (SO_4^{2-}) and nitric acid (HNO_3) gives rise to the nitate ion (NO_3^-). If the acid name ends in –ous, it is modified to end in –ite to name the ion. Sulfurous acid (H_2SO_3) therefore gives rise to the sulfite ion (SO_3^{2-}) and nitrous acid (HNO_2) gives rise to nitrite ion (NO_2^-).

- Acids have many uses, such as:
  - Hydrochloric acid is used to lower the pH of swimming pools.
  - Sulfuric acid is the electrolyte in automobile batteries.
  - Phosphoric acid and carbonic acid are ingredients in soft drinks.

- Citric acid, another organic acid, gives citrus fruit a sour taste.

Common Acids

Set 1 — Common Acids

1. Of the following, which is an acid?

    (1) NaOH(aq)
    (2) NH_3(aq)
    (3) $HC_2H_3O_2$(aq)
    (4) $Ca(OH)_2$(aq)

    1 _____

2. According to the Arrhenius theory, an acid is a substance that

    (1) changes litmus from red to blue
    (2) changes phenolphthalein from colorless to pink
    (3) produces hydronium ions as the only positive ions in an aqueous solution
    (4) produces hydroxide ions as the only negative ions in an aqueous solution

    2 _____

3. Which two formulas represent Arrhenius acids?

    (1) CH_3COOH and CH_3CH_2OH
    (2) $HC_2H_3O_2$ and H_3PO_4
    (3) $KHCO_3$ and $KHSO_4$
    (4) $NaSCN$ and $Na_2S_2O_3$

    3 _____

4. According to one acid-base theory, a water molecule acts as an acid when the water molecule

    (1) accepts an H^+
    (2) accepts an OH^-
    (3) donates an H^+
    (4) donates an OH^-

    4 _____

5. Which substance is an electrolyte?

    (1) CCl_4 (3) HCl
    (2) C_2H_6 (4) H_2O

    5 _____

6. As HCl(g) is added to water, the pH of the water solution

    (1) decreases
    (2) increases
    (3) remains the same

    6 _____

7. What is the pH of a solution that results from the complete neutralization of an HCl solution with a KOH solution?

    (1) 1   (3) 10
    (2) 7   (4) 4

    7 _____

8. Given the following solutions:

    Solution A: pH of 10
    Solution B: pH of 7
    Solution C: pH of 5

    Which list has the solutions placed in order of increasing H^+ concentration?

    (1) A, B, C (3) C, A, B
    (2) B, A, C (4) C, B, A

    8 _____

9. What is the possible pH of a 0.001 M NHO_3?

    (1) 4   (3) 8
    (2) 7   (4) 15

    9 _____

10. Which substance is always a product when an Arrhenius acid in an aqueous solution reacts with an Arrhenius base in an aqueous solution?

    (1) HBr (3) KBr
    (2) H_2O (4) KOH

    10 _____

Base your answers to question 11 using the information below and your knowledge of chemistry.

A beaker contains 100.0 milliliters of a dilute aqueous solution of an acid at equilibrium. The equation below represents this system

$$HC_2H_3O_2(aq) \rightleftharpoons H^+(aq) + C_2H_3O_2^-(aq)$$

11. *a)* Name this acid. __ethanoic acid__

b) Describe what happens to the concentration of $H^+(aq)$ and to the pH when 10 drops of concentrated $HC_2H_3O_2(aq)$ are added to this system.

Concentration: __The concentration of H+ increases__

pH: __decreases__

c) In the equation, the hydronium ion is $H^+(aq)$. Give the other acceptable formula for the hydronium ion. __H_3O^+__

Base your answers to question 12 using the diagrams below and your knowledge of chemistry.

12. Four flasks each contain 100 milliliters of aqueous solutions of equal concentrations at 25°C and 1 atm.

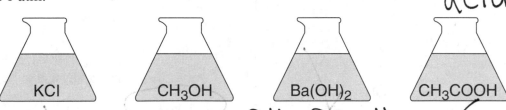

KCl   CH₃OH   Ba(OH)₂   CH₃COOH   acid

a) Which solution is an acidic electrolyte? __CH3COOH__

b) Which solution has the lowest pH? __CH3COOH__

c) What causes aqueous solutions to have a low pH?
__A high concentration of H+ (aq)__

d) Give the formulas of the two beakers that would cause a neutralization reaction.
__CH3COOH and Ba(OH)2__

e) What are the products of a neutralization reaction?
__Water and salt__

f) Which solution contains a salt that is a product of neutralization?
__KCl__

Common Acids

13. Which type of reaction will produce water and a salt?

    (1) saponification (3) esterification
    (2) fermentation (4) neutralization 13 ____

14. Which of these pH numbers indicates the highest level of acidity?

    (1) 5   (3) 10
    (2) 8   (4) 12   14 ____

15. Which technique is safest for diluting a concentrated acid with water?

    (1) add the acid to the water quickly
    (2) add the water to the acid quickly
    (3) add the acid to the water slowly
        while stirring constantly
    (4) add the water to the acid slowly
        while stirring constantly   15 ____

16. A substance that conducts an electrical current when dissolved in water is called

    (1) a calalyst   (3) a nonelectrolyte
    (2) a metalloid   (4) a electrolyte   16 ____

17. Given the equation:

    $HCl(g) + H_2O(\ell) \rightarrow X(aq) + Cl^-(aq)$

    Which ion is represented by X?

    (1) hydroxide   (3) hypochlorite
    (2) hydronium   (4) perchlorate   17 ____

18. Which compound is an Arrhenius acid?
    (1) H_2SO_4 (3) NaOH
    (2) KCl   (4) NH_3 18 ____

19. The compound HNO_3 can be described as an

    (1) Arrhenius acid and an electrolyte
    (2) Arrhenius acid and an nonelectrolyte
    (3) Arrhenius base and an electrolyte
    (4) Arrhenius base and an nonelectrolyte

                                        19 ____

20. One acid-base theory states that an acid is

    (1) an H^- donor   (3) an H^+ donor
    (2) an H^- acceptor   (4) an H^+ acceptor
                                        20 ____

21. Which relationship is present in a solution that has a pH of 4?

    (1) $[H^+] = [OH^-]$ (3) $[H^+] + [OH^-] = 0$
    (2) $[H^+] > [OH^-]$ (4) $[H^+] < [OH^-]$
                                        21 ____

22. Which formula represents a hydronium ion?

    (1) H_3O^+ (3) OH^-
    (2) NH_4^+ (4) HCO_3^- 22 ____

23. Which compound releases hydronium ions in an aqueous solution?

    (1) NaOH   (3) HCl
    (2) CH_3OH (4) KOH   23 ____

24. Which statement describes an alternate theory of acids and bases?

    (1) Acids and bases are both H^+ acceptors.
    (2) Acids and bases are both H^+ donors.
    (3) Acids are H^+ acceptors, and bases
        are H^+ donors.
    (4) Acids are H^+ donors, and bases
        are H^+ acceptors.   24 ____

Base your answers to question 25 using the passage below and your knowledge of chemistry.

Acid rain lowers the pH in ponds and lakes and over time can cause the death of some aquatic life. Acid rain is caused in large part by the burning of fossil fuels in power plants and by gasoline-powered vehicles. The acids commonly associated with acid rain are sulfurous acid, sulfuric acid, and nitric acid. In general, fish can tolerate a pH range between 5 and 9. However, even small changes in pH can significantly affect the solubility and toxicity of common pollutants. Increased concentrations of these pollutants can adversely affect the behavior and normal life processes of fish and cause deformity, lower egg production, and less egg hatching.

25. a) Give the formula for *two* of the acids mention in the article. 1)_____ 2)_____

b) Acid rain caused the pH of a body of water to decrease. Explain this pH decrease in terms of the change in concentration of hydronium ions. _____

_____

c) Give a possible solution to how the acidic lakes pH can be increased. _____

_____

d) Using information in the passage, describe *one* effect of acid rain on future generations of fish species in ponds and lakes. _____

_____

26. A student used blue litmus paper and phenolphthalein as indicators to test the pH of distilled water and five aqueous household solutions. Then the student used a pH meter to measure the pH of the distilled water and each solution. The results of the student's work are recorded in the table.

Testing Results

Liquid Tested	Color of Blue Litmus Paper	Color of Phenolphthalein Paper	Measured pH Value Using a pH Meter
2% milk	blue	colorless	6.4
distilled water	blue	colorless	7.0
household ammonia	blue	pink	11.5
lemon juice	red	colorless	2.3
tomato juice	red	colorless	4.3
vinegar	red	colorless	3.3

a) Identify the liquid tested that has the *lowest* hydronium ion concentration. _____

b) Identify the liquid that is 10 times more acidic than tomato juice. _____

c) Which substance can be used to neutralize vinegar? _____

Common Acids

Answers
Set 1

1. 3   In Table K, $HC_2H_3O_2$(aq) is shown to be ethanoic acid, commonly known as acetic acid.

2. 3   By definition an Arrhenius acid is a compound yielding hydrogen ions (H^+) as the only + ion in aqueous solution. These ions make the solutions acidic.

3. 2   By definition an Arrhenius acid is a compound yielding hydrogen ions (H^+) as the only + ion in aqueous solution. Table K shows that H_3PO_4(aq) is phosphoric acid and $HC_2H_3O_2$(aq) is ethanoic acid (acetic acid).

4. 3   In the Bronsted-Lowry theory, an acid is a substance that donates protons (H^+). When a water molecule donates an H^+ (hydrogen ion), the H^+ will bond to a water molecule, forming a hydronium ion, H_3O^+ (see Table E).

5. 3   An aqueous solution of an acid conducts an electrical current making it an electrolyte. HCl(aq) is hydrochloric acid, and it is an electrolyte.

6. 1   Pure water is neutral with a pH of 7. Acids have a pH less than 7. Therefore, adding HCl(g) to water, producing an aqueous solution of hydrochloric acid, lowers (decreases) the pH of the solution.

7. 2   HCl(aq) is a strong acid and KOH(aq) is a strong base. In the complete neutralization of a strong acid with a strong base, the resulting solution is neutral, having a pH of 7.

8. 1   The lower the pH of a solution, the greater the hydrogen ion (H^+) concentration and the more acidic the solution becomes. Solution C has a higher concentration of hydrogen ions than Solution B, which in turn has a higher concentration of hydrogen ions than Solution A.

9. 1 HNO_3(aq) is nitric acid. All acids, no matter what their strength, will have a pH less than 7.

10. 2   In the neutralization reaction between an acid and a base, the products are a salt and water

11.   Answer: *a*) ethanoic acid *or* acetic acid
              b) concentration of H^+(aq) increases,
                 pH of the system decreases

Explanation: The addition of the concentrated $HC_2H_3O_2$(aq) will shift the equilibrium to the right. This results in an increase in the concentration of H^+(aq). As the H^+(aq) concentration increases, the pH decreases.

c) Answer: H_3O^+

Explanation: The hydrogen ion (a proton) combines with a water molecule to form the hydronium ion, H_3O^+ (also called a hydrated proton).

12. *a*) Answer: $CH_3COOH(aq)$

Explanation: Table K shows that the above formula is that of ethanoic acid. An aqueous solution of an acid conducts an electric current, making it an electrolyte.

b) Answer: $CH_3COOH(aq)$

Explanation: $CH_3COOH(aq)$, being an acid (ethanoic acid), will have a pH less than 7. $Ba(OH)_2$ (aq), being a base, will have a pH greater than 7.
Note: CH_3OH is not a base, but an alcohol possessing the functional group –OH (see Table R).

c) Answer: H^+ ions *or* hydronium ions (H_3O^+)

Explanation: The pH of a solution is a measure of the hydrogen ion (H^+) concentration of the solution. The greater the hydrogen ion concentration, the lower the pH. Acids, having a relatively high concentration of hydrogen ions, have a pH less than 7.

d) Answer: CH_3COOH and $Ba(OH)_2$

Explanation: $Ba(OH)_2$ (a base) will react with $CH_3COOH(aq)$ (ethanoic acid) to produce a salt and water, products of neutralization.
Note: CH_3OH is not a base, but an alcohol possessing the functional group –OH (see Table R).

e) Answer: salt and water

Explanation: Neutralization is the reaction between an acid and a base producing a salt and water.

f) Answer: KCl is potassium chloride, a salt formed from a neutralization reaction.

Common Acids

Table L | Common Bases

Formula	Name
NaOH(aq)	sodium hydroxide
KOH(aq)	potassium hydroxide
$Ca(OH)_2$(aq)	calcium hydroxide
NH_3(aq)	aqueous ammonia

Overview:

Many bases are used as cleaning agents. The household cleaning solution, ammonia, is a base as are oven cleaners and drain cleaners. Strong bases are very dangerous and need to be handled with care. Bases are a class of compounds that possess a characteristic set of physical and chemical properties. Once again, the Swedish chemist Svante Arrhenius proposed a theory on what substances are classified as bases. The Arrhenius theory defines a base as a substance that yields the hydroxide ion (OH^-) as the only negative ion in aqueous solution. This is the ion that gives all aqueous solutions of bases their characteristic properties.

The Table:

This table gives the formula and name of some of the more common bases one encounters in high school chemistry and everyday life. Aqueous ammonia, NH_3(aq), is an aqueous solution of ammonia (NH_3) gas. It may be written as NH_4OH and is called ammonium hydroxide (see Table E). This is the substance in household ammonia cleaning solutions. The strength of the base is not implied by this chart.

Additional Information:

- As a class of compounds, bases exhibit a characteristic set of properties:
  - aqueous solutions of bases conduct an electric current (they are electrolytes)
  - bases neutralize acids (see Table K)
  - bases cause color changes in acid-base indicators (see Table M)
  - aqueous solutions of bases have a slippery feel
  - aqueous solutions of bases have a bitter taste

- Aqueous solutions of bases are also referred to as alkaline solutions.

- Basic or alkaline solutions have a pH greater then 7. The higher the pH, the stronger the base. The stronger the base, the greater the OH^- ion concentration. See page 74, 2nd bullet.

- In the neutralization of a base with an acid, the products are a salt and water.

- The Bronsted-Lowry theory defines a base as a substance that can accept a proton (H^+) from another substance.

- Bases have many uses such as:
  - Sodium hydroxide is an ingredient in oven cleaners and drain cleaners. It is also called lye. It can produce serious chemical burns if spilled on the skin.
  - Potassium hydroxide is used in alkaline batteries.

1. Which pH indicates a basic solution?

    (1) 1   (3) 7 → *neutral*
    (2) 5   (4) 12   1 _____

2. An Arrhenius base yields which ion as the only negative ion in an aqueous solution?

    (1) hydride ion   (3) hydronium ion
    (2) hydrogen ion   (4) hydroxide ion
                                    2 _____

3. Which of these 1 M solutions will have the highest pH?

    (1) NaOH   (3) HCl
    (2) CH_3OH (4) NaCl   3 _____

4. Which substance is an Arrhenius base?

    (1) KCl   (3) KOH
    (2) CH_3Cl (4) CH_3OH 4 _____

acid + base

5. Which compound could serve as a reactant in a neutralization reaction?

    (1) NaCl   (3) CH_3OH
    (2) $Ca(OH)_2$ (4) CH_3CHO 5 _____

6. Which reactants form the salt $CaSO_4(s)$ in a neutralization reaction?

    (1) $H_2S(g)$ and $Ca(ClO_4)_2(s)$
    (2) $H_2SO_3(aq)$ and $Ca(NO_3)_2(aq)$
    (3) $H_2SO_4(aq)$ and $Ca(OH)_2(aq)$
    (4) $SO_2(g)$ and CaO(s)   6 _____

7. Which pair of formulas represents two compounds that are electrolytes?

    (1) HCl and CH_3OH
    (2) HCl and NaOH   A   B
    (3) C_5H_{12} and CH_3OH
    (4) C_5H_{12} and NaOH   7 _____

8. Which relationship is present in a solution that has a pH of 11?

    (1) $[H^+] = [OH^-]$ *Base*
    (2) $[H^+] > [OH^-]$
    (3) $[H^+] < [OH^-]$
    (4) $[H^+] + [OH^-] = 0$ 8 _____

More OH⁻ in base

Base your answers to question 9 using the information below and your knowledge of chemistry.

   Calcium hydroxide is commonly known as agricultural lime and is used to adjust the soil pH. Before the lime was added to a field, the soil pH was 5. After the lime was added, the soil underwent a 100-fold decrease in hydronium ion concentration.

9. *a)* What ion caused the decrease of hydronium ion? ___ OH^- ___

   b) Give the formula for the base used in this passage. ___ $Ca(OH)_2$ (aq) ___

10. According to the Arrhenius theory, a base reacts with an acid to produce

    (1) ammonia and methane
    (2) ammonia and a salt
    (3) water and methane
    (4) water and a salt   10 ____

11. Which compound could neutralize HCl(aq)?

    (1) $Ca(OH)_2$(aq)   (3) NaCl
    (2) H_2O (4) H_2SO_3(aq)   11 ____

12. According to the Arrhenius theory, when a base dissolves in water it produces

    (1) CO_3^{2-} as the only negative ion in solution
    (2) OH^- as the only negative ion in solution
    (3) NH_4^+ as the only positive ion in solution
    (4) H^+ as the only positive ion in solution   12 ____

13. Equal volumes of 0.1 M NaOH and 0.1 M HCl are thoroughly mixed. The resulting solution has a pH closest to

    (1) 5   (3) 3
    (2) 7   (4) 9   13 ____

14. As an acid solution is added to neutralize a base solution, the OH^- concentration of the base solution

    (1) decreases
    (2) increases
    (3) remains the same   14 ____

15. Given the following solutions:

    Solution A: pH of 10
    Solution B: pH of 8
    Solution C: pH of 7

    Which list has the solutions placed in order of increasing OH^- concentration?

    (1) A, B, C (3) C, A, B
    (2) B, A, C (4) C, B, A 15 ____

16. Which compound releases hydroxide ions in an aqueous solution?

    (1) CH_3COOH (3) HCl
    (2) CH_3OH (4) KOH   16 ____

17. Given the reaction: $NH_3 + HCl \rightarrow NH_4Cl$

    In this reaction, ammonia molecules (NH_3) act as a base because they

    (1) accept hydrogen ions (H^+)
    (2) donate hydrogen ions (H^+)
    (3) accept hydroxide ions (OH^-)
    (4) donate hydroxide ions (OH^-)   17 ____

18. Which salt is produced when sulfuric acid and calcium hydroxide react completely?

    (1) CaH_2 (3) CaS
    (2) CaO   (4) $CaSO_4$ 18 ____

19. Which chemical equation represents the reaction of an Arrhenius acid and an Arrhenius base?

    (1) $HC_2H_3O_2$(aq) + NaOH(aq) $\rightarrow$ $NaC_2H_3O_2$(aq) + H_2O(ℓ)
    (2) C_3H_8(g) + $5O_2$(g) $\rightarrow$ 3 CO_2(g) + $4H_2O$(ℓ)
    (3) Zn(s) + 2 HCl(aq) $\rightarrow$ $ZnCl_2$(aq) + H_2(g)
    (4) $BaCl_2$(aq) + Na_2SO_4(aq) $\rightarrow$ $BaSO_4$(s) + 2NaCl(aq)   19 ____

Base your answers to question 20 using the information below and your knowledge of chemistry.

Three bottles of liquids labeled 1, 2, and 3 were found in a storeroom. One of the liquids is known to be drain cleaner. Drain cleaners commonly contain KOH or NaOH. The pH of each liquid at 25°C was determined with a pH meter. The table below shows the test results.

pH Test Results

Bottle	pH of Liquid
1	3.8
2	7.0
3	12.8

20. *a)* Explain how the pH results in this table enable a student to correctly conclude that bottle 3 contains the drain cleaner.

_____

_____

b) Which bottle would have the highest concentration of OH⁻ ions? _____

c) Which bottle could contain distilled water? _____

d) Liquid from bottle 1 is gradually added to bottle 3. Explain what happens to the pH of the liquid in bottle 3.

_____

Base your answers to question 21 using the information below and your knowledge of chemistry.

A student was studying the pH differences in two samples of liquid waste. The student measured a pH of 9 in container *A* and a pH of 12 in container *B*.

21. *a)* Compare the hydroxide ion concentration in container *A* to the hydroxide ion concentration in container *B*.

_____

_____

b) Explain why mixing container *A* and container *B* will not produce neutralization.

_____

_____

c) Identify one compound that could be used to neutralize sample *B*. _____

Table L – Common Bases
Answers – Set 1

1. 4   A solution with a pH of 7 is considered neutral. A basic solution has a pH greater than 7.

2. 4   By definition, an Arrhenius base is a compound that yields hydroxides ions (OH^-) as the only negative ion in aqueous solution.

3. 1   Bases have pH values greater than 7. Sodium hydroxide is the only base of the given choices (see Table L). CH_3OH is an alcohol (see Table R – alcohols) and does not ionize to produce a basic solution.

4. 3   By definition, an Arrhenius base is a substance that yields hydroxide ion (OH^-) as the only negative ion in an aqueous solution. KOH(aq) is a base as shown in Table L. It will ionize in water, producing OH^- ions. CH_3OH is an alcohol (see Table R – alcohols) and does not ionize in water to yield OH^- ions.

5. 2   In a neutralization reaction, an acid and a base react to produce a salt and water. Table L shows that $Ca(OH)_2$(aq), calcium hydroxide, is a base (contains the OH^- ion) and could be the reactant in a neutralization reaction. Choice 3 is wrong since it is an organic alcohol containing the functional group –OH, see Table R.

6. 3   In a neutralization reaction, an acid and a base react producing a salt and water. In Table K, H_2SO_4(aq) is shown to be sulfuric acid and in Table L, $Ca(OH)_2$(aq) is shown to be calcium hydroxide, a base. These reactants would produce $CaSO_4$ (a salt) and H_2O.

7. 2   A substance that conducts an electrical current when dissolved in water is called an electrolyte. Examples of some electrolytes are all ionic compounds (salts), bases and acids. Open to Table K and L and locate hydrochloric acid and sodium hydroxide, respectively. Being an acid and a base, they are electrolytes. CH_3OH in choice 3 is an alcohol, (identified by the functional group –OH, see Table R) and does not ionize, thus will not conduct an electrical current. Choices 3 and 4 show a hydrocarbon molecule, C_5H_{12}. These do not conduct electricity.

8. 3   Basic solutions have a pH greater than 7. All bases therefore have a greater concentration of hydroxide ions (OH^-) than hydronium ions (H_3O^+).

9. *a*)   Answer: OH^- *or* hydroxide

   Explanation: Calcium hydroxide is a base (see Table L). The soil, with a pH of 5, is acidic. The OH^- ions from the base neutralizes the excess hydronium ions in the soil.

   b)   Answer: $Ca(OH)_2$(aq)

   Explanation: The equation for calcium hydroxide is shown in Table L.

Indicator	Approximate pH Range for Color Change	Color Change
methyl orange	3.1–4.4	red to yellow
bromthymol blue	6.0–7.6	yellow to blue
phenolphthalein	8–9	colorless to pink
litmus	4.5–8.3	red to blue
bromcresol green	3.8–5.4	yellow to blue
thymol blue	8.0–9.6	yellow to blue

Source: *The Merck Index*, 14th ed., 2006, Merck Publishing Group

Overview:

Indicators are substances that show different colors in acid and base solutions. Therefore, they can be used to "indicate" whether a solution is an acid or a base. The color of an indicator is sensitive to the pH of the solution. An indicator changes color at a particular pH or pH range.

The Table:

The table lists some common indicators, the approximate pH range for color change and the color change. Usually, the indicator changes color gradually from the color shown at the low pH to the color shown at the high pH. One of the given indicators is litmus. Litmus may come in the form of a liquid or paper. Red litmus paper turns blue in a basic solution. Blue litmus paper turns red in an acidic solution.

Additional Information:

- An indicator can be used to determine the end point in an acid-base titration. Titration is the gradual addition of an acid to a base, or vice versa, until neutralization occurs. At the endpoint, the indicator changes color. At the endpoint, the resulting solution is not necessarily neutral (pH = 7). It depends upon the strength of the acid and base.

- Since pH is a logarithmic function, a change in value of one pH unit represents a 10 fold change in hydrogen ion concentration. For example, a decrease in pH of one unit, indicating that the solution becomes more acidic (see Table K) and that the hydrogen ion (H^+) concentration is ten times greater than the original concentration. If the pH increases by 3 units, indicating that the solution becomes more basic (see Table L), the hydrogen ion concentration is 1/1000 (1/10 × 1/10 × 1/10) the original concentration.

- Phenolphthalein is a very common indicator used in chemistry labs to show the endpoint of a titration.

- Some indicator substances are obtained from vegetables. The juice left from boiling red cabbage can be used as an indicator.

1. In which 0.01 M solution is phenolphthalein pink?

   (1) $CH_3OH(aq)$ (3) $CH_3COOH(aq)$
   (2) $Ca(OH)_2(aq)$ (4) $HNO_3(aq)$ 1 _____

2. In which solution will thymol blue indicator appear blue?

   (1) 0.1 M CH_3COOH
   (2) 0.1 M KOH
   (3) 0.1 M HCl
   (4) 0.1 M H_2SO_4 2 _____

3. Which indicator, when added to a solution, changes color from yellow to blue as the pH of the solution is changed from 5.5 to 8.0?

   (1) bromcresol green (3) litmus
   (2) bromthymol blue (4) methyl orange

   3 _____

4. Which indicator would best distinguish between a solution with a pH of 3.5 and a solution with a pH of 5.5?

   (1) bromthymol blue (3) litmus
   (2) bromcresol green (4) thymol blue

   4 _____

5. As the pH of a solution is changed from 3 to 6, the concentration of hydronium ions

   (1) increases by a factor of 3
   (2) increases by a factor of 1000
   (3) decreases by a factor of 3
   (4) decreases by a factor of 1000   5 _____

6. A student was given four unknown solutions. Each solution was checked for conductivity and tested with phenolphthalein. The results are shown in the data table below.

Solution	Conductivity	Color with Phenolphthalein
A	Good	Colorless
B	Poor	Colorless
C	Good	Pink
D	Poor	Pink

   Based on the data table, which unknown solution could be 0.1 M NaOH?

   (1) A   (3) C
   (2) B   (4) D   6 _____

7. A student tested a 0.1 M aqueous solution and made the following observations:

   • conducts electricity
   • turns blue litmus to red
   • reacts with Zn(s) to produce gas bubbles

   Which compound could be the solute in this solution?

   (1) CH_3OH (3) HCl
   (2) LiBr   (4) LiOH   7 _____

8. Which statement describes the characteristics of an Arrhenius base?

   (1) It changes blue litmus to red and has a pH less than 7.
   (2) It changes blue litmus to red and has a pH greater than 7.
   (3) It changes red litmus to blue and has a pH less than 7.
   (4) It changes red litmus to blue and has a pH greater than 7.   8 _____

Base your answers to question 9 using the passage below and your knowledge of chemistry.

Acid rain is a problem in industrialized countries around the world. Oxides of sulfur and nitrogen are formed when various fuels are burned. These oxides dissolve in atmospheric water droplets that fall to earth as acid rain or acid snow.

While normal rain has a pH between 5.0 and 6.0 due to the presence of dissolved carbon dioxide, acid rain often has a pH of 4.0 or lower. This level of acidity can damage trees and plants, leach minerals from the soil, and cause the death of aquatic animals and plants. If the pH of the soil is too low, then quicklime, CaO, can be added to the soil to increase the pH. Quicklime produces calcium hydroxide when it dissolves in water.

9. *a)* A sample of wet soil has a pH of 4.0. After the addition of quicklime, the H^+ ion concentration of the soil is 1/100 of the original H^+ ion concentration of the soil. What is the new pH of the soil sample? _____

b) Identify *one* indicator from Reference Table M that is yellow in a solution with a pH of 4.0. _____

c) Identify an indicator, that can be used to test acid rain having a pH of 5.1. _____

Base your answers to question 10 using the information below and your knowledge of chemistry.

Sulfur dioxide, SO_2, is one gas produced when fossil fuels are burned. When this gas reacts with water in the atmosphere, an acid is produced forming acid rain. The pH of the water in a lake changes when acid rain collects in the lake.

Two samples of this rainwater are tested using two indicators. Methyl orange is yellow in one sample of this rainwater. Litmus is red in the other sample of this rainwater.

10. *a)* Identify a possible pH value for the rainwater that was tested. _____

b) Write the formula for one substance that can neutralize the lake water affected by acid rain.

_____

c) Give the formula for the acid that is formed when SO_2 dissolves with water.

_____

d) Give the name of the acid that is formed when SO_2 dissolves with water.

_____

Common Acid – Base Indicators

11. Which solution when mixed with a drop of bromthymol blue will cause the indicator to change from blue to yellow?

    (1) 0.1 M HCl   (3) 0.1 M CH$_3$OH
    (2) 0.1 M NH$_3$ (4) 0.1 M NaOH

    11 _____

12. Which indicator is blue in a solution that has a pH of 5.6?

    (1) bromcresol green   (3) methyl orange
    (2) bromthymol blue   (4) thymol blue

    12 _____

13. A compound whose water solution conducts electricity and turns phenolphthalein pink is

    (1) HCl   (3) NaOH
    (2) HC$_2$H$_3$O$_2$ (4) CH$_3$OH   13 _____

14. Which statement correctly describes a solution with a pH of 9?

    (1) It has a higher concentration of H$_3$O$^+$ than OH$^-$ and causes litmus to turn blue.
    (2) It has a higher concentration of OH$^-$ than H$_3$O$^+$ and causes litmus to turn blue.
    (3) It has a higher concentration of H$_3$O$^+$ than OH$^-$ and causes methyl orange to turn yellow.
    (4) It has a higher concentration of OH$^-$ than H$_3$O$^+$ and causes methyl orange to turn red.   14 _____

15. Which pH change represents a hundredfold increase in the concentration of H$_3$O$^+$?

    (1) pH 5 to pH 7   (3) pH 3 to pH 1
    (2) pH 13 to pH 14   (4) pH 4 to pH 3

    15 _____

16. According to Reference Table M, what is the color of the indicator methyl orange in a solution that has a pH of 2?

    (1) blue   (3) orange
    (2) yellow   (4) red   16 _____

17. A solution with a pH of 11 is first tested with phenolphthalein and then with litmus. What is the color of each indicator in this solution?

    (1) Phenolphthalein is colorless and litmus is blue.
    (2) Phenolphthalein is colorless and litmus is red.
    (3) Phenolphthalein is pink and litmus is blue.
    (4) Phenolphthalein is pink and litmus is red.   17 _____

18. Based on the results of testing colorless solutions with indicators, which solution is most acidic?

    (1) a solution in which bromthymol blue is blue
    (2) a solution in which bromcresol green is blue
    (3) a solution in which phenolphthalein is pink
    (4) a solution in which methyl orange is red

    18 _____

19. The pH of an aqueous solution changes from 4 to 3 when the hydrogen ion concentration in the solution is

    (1) decreased by a factor of $\frac{3}{4}$
    (2) decreased by a factor of 10
    (3) increased by a factor of $\frac{4}{3}$
    (4) increased by a factor of 10   19 _____

20. A student is given two beakers, each containing an equal amount of clear, odorless liquid. One solution is acidic and the other is basic.

a) State *two* safe methods of distinguishing the acid solution from the base solution.

Method one: _____

_____

Method two: _____

_____

b) For *each* method, state the results of both the testing of the acid solution *and* the testing of the base solution.

Results of method one: _____

_____

Results of method two: _____

_____

Base your answers to question 21 on the information below.

A student, wearing chemical safety goggles and a lab apron, is to perform a laboratory test to determine the pH value of two different solutions. The student is given one bottle containing a solution with a pH of 2.0 and another bottle containing a solution with a pH of 5.0. The student is also given six dropping bottles, each containing a different indicator listed in Reference Table M.

21. a) State one safety precaution, not mentioned in the passage, that the student should take while performing tests on the samples from the bottles.

_____

b) Identify an indicator in Reference Table M that would differentiate the two solutions.

_____

c) Compare the hydronium ion concentration of the solution having a pH of 2.0 to the hydronium ion concentration of the other solution given to the student.

_____

d) What color would the solution having a pH of 5.0 be when bromcresol green is added?

_____

Common Acid – Base Indicators

Base your answers to question 22 using the information below and your knowledge of chemistry.

Calcium hydroxide is commonly known as agricultural lime and is used to adjust the soil pH. Before the lime was added to a field, the soil pH was 5. After the lime was added, the soil underwent a 100-fold decrease in hydronium ion concentration.

22. *a)* What is the new pH of the soil in the field? _____

b) What indicator could the student use to test the pH of the soil after calcium hydroxide was added to the soil? _____

c) Give the chemical symbol of the ion that decreased in concentration. _____

Base your answers to question 23 on the information below.

A person experiencing acid indigestion after drinking tomato juice can ingest milk of magnesia to reduce the acidity of the stomach contents. Tomato juice has a pH value of 4. Milk of magnesia, a mixture of magnesium hydroxide and water, has a pH value of 10.

23. *a)* Compare the hydrogen ion concentration in tomato juice to the hydrogen ion concentration in milk of magnesia.

_____

b) Identify the negative ion found in milk of magnesia. _____

c) What is the color of thymol blue indicator when placed in a sample of milk of magnesia?

_____

Base your answers to question 24 on the information below.

Some carbonated beverages are made by forcing carbon dioxide gas into a beverage solution. When a bottle of one kind of carbonated beverage is first opened, the beverage has a pH value of 3.

24. *a)* State, in terms of the pH scale, why this beverage is classified as acidic.

_____

b) Using Table M, identify one indicator that is yellow in a solution that has the same pH value as this beverage. _____

c) After the beverage bottle is left open for several hours, the hydronium ion concentration in the beverage solution decreases to 1/1000 of the original concentration. Determine the new pH of the beverage solution. _____

1. 2   Phenolphthalein will change from colorless to pink over a pH range of 8 to 9. This pH range indicates the solution is basic. The only base given is $Ca(OH)_2(aq)$, calcium hydroxide. Choice 1 is an alcohol shown by the functional group –OH (see Table R under alcohol).

2. 2   Table M shows that thymol blue will change from yellow to blue over a pH range of 8.0 to 9.6. This indicates the solution is basic. Table L shows that KOH(aq) is a base. All other answers are acids as shown in Table K and would have a pH lower than 7.

3. 2   Table M gives different indicators and their color change range. Bromthymol blue will undergo a color change from yellow to blue as the pH of the solution is changed from 6.0 to 7.6.

4. 2   Table M shows that bromcresol green will change from yellow to blue over a pH range of 3.8 – 5.4. A solution containing this indicator would be yellow at a pH of 3.5 and blue at a pH of 5.5.

5. 4   A decrease in hydronium ions causes the pH to increase. The pH is a logarithmic function. A change in value of one pH unit represents a 10-fold change in hydrogen ion concentration; a change in value of 2 pH units represents a 100-fold change and a change in value of 3 pH units represents a 1000-fold change. Since the pH is increasing, the hydrogen ion concentration will be decreasing.

6. 3   Referring to Table L, it shows that NaOH(aq) is a base. Bases are electrolytes (conductors of electricity) since they ionize in aqueous solution. Bases will have pH values higher than 7. This will cause phenolphthalein to be pink.

7. 3   Acids turn blue litmus paper red and are considered electrolytes since they conduct electricity. Also, acids react with certain metals (those above H_2 on Table J), such as zinc, to produce bubbles of hydrogen gas. HCl dissolves in water producing hydrogen ions (H^+), causing the solution to be acidic. Answer 1 is an alcohol shown by the functional group –OH (see Table R), answer 2 is lithium bromide a salt, and answer 4 is lithium hydroxide, a base, shown by the hydroxide ion (OH^-).

8. 4   All bases will have a pH greater than 7. Litmus paper is red in an acidic solution and blue in a basic solution.

9. *a)* Answer: 6.0 *or* 6

Explanation: The pH scale indicates the concentration of the hydrogen ions (H^+) in a solution. The pH scale is a logarithmic function. The wet soil was at a pH of 4. After quicklime was added to the soil, the soil was 1/100 less acidic. Since the soil was at a pH of 4, a pH change from 4 to 5 represents a 1/10 decrease in H^+ or acidity, and a change from 5 to 6 would represent an additional 1/10 decrease in acidity, making the total change of 1/100 (1/10 × 1/10 = 1/100). Thus, the new pH must be 6.

b) Answer: methyl orange

Explanation: Table M shows that methyl orange would be yellow in a solution with a pH of 4.

c) Answer: bromcresol green

Explanation: Table M shows that bromcresol green undergoes a color change in the pH range of 3.8 to 5.4.

10. *a)* Answer: An acceptable answer would be any pH value from 4.0 to 5.0

Explanation: Methyl orange has a pH range of 3.1 to 4.4, in which the color gradually changes from red to yellow. The sample is yellow with methyl orange, so its pH is close to 4.4 or higher. Litmus has a pH range for color change from 4.5 to 8.3, in which litmus changes from red to blue. Litmus is red so the sample's pH must close to 4.5. Since pH indicators change color over a range, the acceptable pH range for this sample would be from 4.0 to 5.0.

b) Answer: $Ca(OH)_2(aq)$ *or* $KOH(aq)$ *or* $NH_3(aq)$ *or* $NaOH(aq)$

Explanation: Bases neutralize acids. A base contains hydroxide ions (OH^-) that react with the hydronium ions (H_3O^+) in the lake water, neutralizing the effect of the acid rain. The above answers are all bases.

c) Answer: $H_2SO_3(aq)$

Explanation: When SO_2 dissolved in water $SO_2 + H_2O \rightarrow H_2SO_3(aq)$ is produced. (See Table K – Common Acid)

d) Answer: Sulfurous acid

Explanation: See the 6th bullet on page 62.

Nuclide	Half-Life	Decay Mode	Nuclide Name
^{198}Au	2.695 d	β^-	gold-198
^{14}C	5715 y	β^-	carbon-14
^{37}Ca	182 ms	β^+	calcium-37
^{60}Co	5.271 y	β^-	cobalt-60
^{137}Cs	30.2 y	β^-	cesium-137
^{53}Fe	8.51 min	β^+	iron-53
^{220}Fr	27.4 s	α	francium-220
^{3}H	12.31 y	β^-	hydrogen-3
^{131}I	8.021 d	β^-	iodine-131
^{37}K	1.23 s	β^+	potassium-37
^{42}K	12.36 h	β^-	potassium-42
^{85}Kr	10.73 y	β^-	krypton-85
^{16}N	7.13 s	β^-	nitrogen-16
^{19}Ne	17.22 s	β^+	neon-19
^{32}P	14.28 d	β^-	phosphorus-32
^{239}Pu	2.410×10^4 y	α	plutonium-239
^{226}Ra	1599 y	α	radium-226
^{222}Rn	3.823 d	α	radon-222
^{90}Sr	29.1 y	β^-	strontium-90
^{99}Tc	2.13×10^5 y	β^-	technetium-99
^{232}Th	1.40×10^{10} y	α	thorium-232
^{233}U	1.592×10^5 y	α	uranium-233
^{235}U	7.04×10^8 y	α	uranium-235
^{238}U	4.47×10^9 y	α	uranium-238

Source: *CRC Handbook of Chemistry and Physics*, 91st ed., 2010–2011, CRC Press

Overview:

The nucleus of many elements is unstable and gives off particles and/or energy when going to a more stable state. These elements are called radioactive elements. This process is referred to as radioactive decay and the emitted particles or energy is referred to as radiation. The term radioisotope is a contraction of the words radioactive isotope.

The Table:

Table N lists the Nuclide, Half-Life, Decay Mode and the Nuclide Name of selected radioisotopes. The term nuclide refers to any nucleus. It is in the nucleus where the decay of the radioisotopes occurs. In the first column, the symbol and mass number are given for each nuclide. When needed, the atomic number for any of these nuclides is obtained from the Periodic Table. The half-life of a radioisotope is the time in which one-half the nuclei of a sample of that radioisotope decays. The decay mode indicates what particle is emitted as the nucleus undergoes decay. The decay modes listed on the table are:

β^- - negative beta decay, which is the emission of an ordinary electron.

β^+ - positive beta decay, which is the emission of a positive electron or a positron.

α - alpha decay, which is the emission of a particle identical to a helium nucleus.

Note: See Table O for Symbol and Notation.

Radioactive Decay:

During radioactive decay, a radioactive element gradually changes into another element. The time in which one-half life of the nuclei of the original element decays is called the half-life. Table N gives the half-lives of selected radioisotopes. Each radioactive isotope has its own half-life that is unaffected by pressure, temperature or any other external factors. When a substance undergoes a radioactive decay, the radiation decreases, but the half-life remains constant. At the conclusion of each half-life, the mass of the radioactive sample is one half-life of the mass it had at the beginning of that half-life.

Example 1: What amount remains of a 20 gram sample of radium-226 remains after 4,797 years?

Solution: From Table N, the half-life of radium-226 is 1,599 years.

Since the half-life is constant, 4,797 y equals 3 half-lives $\left(\dfrac{4797\,y}{1599\,y} \right)$.

After each half-life, the amount of radioactive isotope remaining would be half of the original amount, or:

- after the 1st half-life, 10 grams remain,
- after the 2nd half-life, 5 grams remain and
- after the 3rd half-life, 2.5 grams remain.

Answer: 2.5 g

Example 2: What fraction remains of a 20 gram sample of radium-226 remains after 4,797 years?

This problem is the same as *Example 1*, except that it is asking for the fraction that remains after 4,797 years.

Solution: Using the same procedure as *Example 1* and including the fractional amounts we have:

- after the 1ˢᵗ half-life, 10 grams remain, which is 1/2 of the original amount,
- after the 2ⁿᵈ half-life, 5 grams remain, which is 1/4 of the original amount and
- after the 3ʳᵈ half-life, 2.5 grams remain, which is 1/8 of the original amount.

Answer: 1/8

Example 3: How much time must elapse before 16 grams of potassium-42 decays, leaving 2 grams of the original isotope?

Solution: Open to Table N. It shows that potassium-42 (^{42}K) has a half-life of 12.36 hours. This is the time it takes for half of the radioactive substance to decay. Since the half-life is a constant, the time for the first half-life would be 1×12.36 h or 12.36 hours and the amount remaining would be half of the original or 8 grams. The time for two half-lives would be 2×12.36 h or 24.72 hours and 4 grams would remain. The time for three half-lives would be 3×12.36 h or 37.08 hours and 2 grams of the original sample would remain radioactive.

Answer: 37.08 h

Additional Information:

- In most cases, the identity (name) of the nuclide changes during radioactive decay.

- Another decay mode not shown on the table is gamma emission (γ), which is the emission of pure energy from the nucleus.

- Radioisotopes are used in dating geologic and archeological finds. C-14, found in all plant and animal matter, is used to date such finds. U-238 is used to date minerals.

- Radioisotopes with short half-lives, which will be quickly eliminated from the body are used medicinally. An example is the use of I-131, with a half-life of 8.021 days, to treat thyroid disorders. Other medical uses of radioisotopes are:
  - Ra-226 and Co-60 are used in cancer therapy.
  - Tc-99 is used to pinpoint brain tumors.

- Radiation can be used to kill insect eggs, molds, bacteria and yeasts in foods, increasing the shelf-life of certain food items.

1. According to Reference Table N, which pair of isotopes spontaneously decays?

   (1) C-12 and N-14
   (2) C-12 and N-16
   (3) C-14 and N-14
   (4) C-14 and N-16   1 _____

2. Alpha particles are emitted during the radioactive decay of

   (1) carbon-14
   (2) neon-19
   (3) calcium-37
   (4) plutonium-239   2 _____

3. What is the half-life and decay mode of Rn-222?

   (1) 1.91 days and alpha decay
   (2) 1.91 days and beta decay
   (3) 3.823 days and alpha decay
   (4) 3.823 days and beta decay   3 _____

4. As a sample of the radioactive isotope ^{131}I decays, its half-life

   (1) decreases
   (2) increases
   (3) remains the same   4 _____

5. Which nuclide has a half-life that is less than one minute?

   (1) cesium-137
   (2) francium-220
   (3) phosphorus-32
   (4) strontium-90   5 _____

6. Which isotope is most commonly used in the radioactive dating of the remains of organic materials?

   (1) ^{14}C   (3) ^{32}P
   (2) ^{16}N   (4) ^{37}K   6 _____

7. Which equation represents the radioactive decay of $^{226}_{88}$Ra ?

   (1) $^{226}_{88}\text{Ra} \rightarrow {}^{222}_{86}\text{Rn} + {}^{4}_{2}\text{He}$
   (2) $^{226}_{88}\text{Ra} \rightarrow {}^{222}_{89}\text{Ac} + {}^{0}_{-1}\text{e}$
   (3) $^{226}_{88}\text{Ra} \rightarrow {}^{222}_{87}\text{Fr} + {}^{0}_{+1}\text{e}$
   (4) $^{226}_{88}\text{Ra} \rightarrow {}^{225}_{88}\text{Ra} + {}^{0}_{1}\text{n}$ 7 _____

8. Which radioisotope is used in medicine to treat thyroid disorders?

   (1) cobalt-60   (3) phosphorus-32
   (2) iodine-131   (4) uranium-238   8 _____

9. Which notation of a radioisotope is correctly paired with the notation of its emission particle?

   (1) ^{37}Ca and ${}^{4}_{2}$He   (3) ^{16}N and ${}^{1}_{1}$p
   (2) ^{235}U and ${}^{0}_{+1}$e   (4) ^{3}H and ${}^{0}_{-1}$e   9 _____

10. Based on Reference Table N, what fraction of a radioactive ^{90}Sr sample would remain unchanged after 58.2 years?

    (1) $\frac{1}{2}$ (3) $\frac{1}{8}$
    (2) $\frac{1}{4}$ (4) $\frac{1}{16}$ 10 _____

11. An article states "Within the 10,000-year time period, cesium and strontium, the most powerful radioactive emitters, would have decayed." Use information from Reference Table N to support this statement.

_____

_____

12. How is the radioactive decay of krypton-85 different from the radioactive decay of francium-220?

_____

_____

Base your answers to question 13 using the passage below and your knowledge of chemistry.

In living organisms, the ratio of the naturally occurring isotopes of carbon, C-12 to C-13 to C-14, is fairly consistent. When an organism such as a woolly mammoth died, it stopped taking in carbon, and the amount of C-14 present in the mammoth began to decrease. For example, one fossil of a woolly mammoth is found to have 1/32 of the amount of C-14 found in a living organism.

13. *a*) Identify the type of nuclear reaction that caused the amount of C-14 in the woolly mammoth to decrease after the organism died.

_____

b) State in terms of subatomic particles, how an atom of C-13 is different from an atom of C-12.

_____

_____

c) What is the half-life of C-14? _____

14. The radioisotopes carbon-14 and nitrogen-16 are present in a living organism. Carbon-14 is commonly used to date a once-living organism.

A sample of wood is found to contain $\frac{1}{8}$ as much C-14 as is present in the wood of a living tree. What is the approximate age, in years, of this sample of wood? _____ y

15. If a scientist purifies 10.0 gram of radium-226, how many years must pass before only 0.625 gram of the original radium-226 sample remains unchanged? _____ y

16. Which isotope is radioactive?

    (1) C-12   (3) Tc-99
    (2) Ne-20   (4) Pb-206   16 _____

17. What is the decay mode of ^{37}K?

    (1) β⁻   (3) γ
    (2) β⁺   (4) α   17 _____

18. When cobalt-60 undergoes nuclear decay, it emits

    (1) a positron
    (2) a neutron
    (3) a beta particle
    (4) an alpha particle   18 _____

19. According to Table N, which radioactive isotope is best for determining the actual age of Earth?

    (1) ^{238}U (3) ^{60}Co
    (2) ^{90}Sr (4) ^{14}C 19 _____

20. Which radioisotope undergoes beta decay and has a half-life of less than 1 minute?

    (1) Fr-220   (3) N-16
    (2) K-42   (4) P-32   20 _____

21. Which two radioisotopes have the same decay mode?

    (1) ^{37}Ca and ^{53}Fe
    (2) ^{220}Fr and ^{60}Co
    (3) ^{37}K and ^{42}K
    (4) ^{99}Tc and ^{19}Ne 21 _____

22. Which nuclides are used to date the remains of a once-living organism?

    (1) C-14 and C-12
    (2) I-131 and Xe-131
    (3) Co-60 and Co-59
    (4) U-238 and Pb-206   22 _____

23. The isotopes K-37 and K-42 have the same

    (1) total number of neutrons in their atoms
    (2) bright-line spectrum
    (3) mass number for their atoms
    (4) decay mode   23 _____

24. Which nuclide is listed with its half-life and decay mode?

    (1) K-37, 1.23 h, α
    (2) N-16, 7.13 s, β⁻
    (3) Rn-222, 1.599×10^3 y, α
    (4) U-235, 7.04×10^8 y, β⁻   24 _____

25. Which radioisotope has an atom that emits a particle with a mass number of 0 and a charge of +1?

    (1) ^3H (3) ^{19}Ne
    (2) ^{16}N (4) ^{239}Pu 25 _____

26. Atoms of I-131 spontaneously decay when the

    (1) stable nuclei emit alpha particles
    (2) stable nuclei emit beta particles
    (3) unstable nuclei emit alpha particles
    (4) unstable nuclei emit beta particles

        26 _____

27. According to Reference Table N, which radioactive isotope will retain only one-eighth ($\frac{1}{8}$) its original radioactive atoms after approximately 42.84 days?

(1) gold-198
(2) iodine-131
(3) phosphorus-32
(4) radon-222   27 ____

28. Which radioisotope emits alpha particles?

(1) Fe-53   (3) Au-198
(2) Sr-90   (4) Pu-239   28 ____

29. The nucleus of a radium-226 atom is unstable, which causes the nucleus to spontaneously

(1) absorb electrons   (3) decay
(2) absorb protons   (4) oxidize   29 ____

30. The decay of N-16 is represented by the balanced equation: $^{16}_{7}N \rightarrow ^{0}_{-1}e + ^{16}_{8}O$

What is the decay mode for nitrogen-16? ____ β⁻ ____

31. Cesium-137 is sometimes used in radiation therapy as a treatment for cancer. A sample of cesium-137 was left in an abandoned clinic in Brazil in 1987. Cesium-137 gives off a blue glow because of its radioactivity. The people who discovered the sample were attracted by the blue glow and had no idea of any danger. Hundreds of people were treated for overexposure to radiation, and four people died.

Suppose a 40-gram sample of iodine-131 and a 40-gram sample of cesium-137 were both abandoned in a lab in 1987. Explain why the sample of iodine-131 would not pose as great a radiation risk to people today as the sample of cesium-137 would.

I-131 has a half life of 8.031 days so it wouldn't be radioactive. CS-131 has a half life of 30.2 years making it more radioactive today.

32. A 400-gram sample of strontium-90 undergoes radioactive decay. For each half-life that is listed below, give the percentage that remains, the mass in grams that remains, and the number of years that have past for the original sample.

1st half-life - ___400___ grams, ___50___%, ___29.1___ years past

2nd half-life - ___200___ grams, ___$\frac{1}{2}$ 25___%, ___58.2___ years past

3rd half-life - ___100___ grams, ___$\frac{1}{4}$___%, ___87.3___ years past

4th half-life - ___50___ grams, _____%, ___116.4___ years past

33. Based on Reference Table N, what is the fraction of a sample of potassium-42 that will remain unchanged after 61.8 hours? ____ 1/32 ____

61.8 ÷ 12.36

1. 4 Isotopes that undergo spontaneous decay are radioactive elements. Only choice 4 shows a pair of elements that are found in Table N – Selected Radioisotopes.

2. 4 Referring to Table O, the alpha particle is represented by the symbol α. In Table N, it shows that Pu-239 decays by emitting an alpha particle.

3. 3 In Table N locate the radioisotope Rn-222. The given half-life is 3.823 days and its decay mode is alpha particle emission (α).

4. 3 Half-life is the time it takes for one half of a radioactive element to decay. The half-life for a radioactive element is a constant. Thus in time, the original radioactive substance decreases in amount, but the half-life remains the same.

5. 2 From Table N, the only radioisotope of the given choices that has a half-life less than one minute is ^{220}Fr with a half-life of 27.4 s.

6. 1 Carbon is found in all organic matter. Most of this carbon is ^{12}C, while some of it is radioactive ^{14}C. Table N shows that ^{14}C has a half-life of 5715 years. Due to the relatively long half-life, ^{14}C has been used extensively in dating organic materials.

7. 1 Open to Table N and in the Decay Mode, it shows that Ra-226 will emit an alpha particle when it undergoes radioactive decay. The release of an alpha particle, being a helium nucleus ($^{4}_{2}He$), reduces the atomic mass of the radioactive element by 4 and its atomic number by 2. Choice 1 shows the proper decay mode and a nuclear equation conserving mass and charge.

8. 2 The thyroid gland readily absorbs iodine. For many thyroid disorders, doctors have the patient drink an "atomic cocktail" containing radioactive iodine. Open to Table N, and it shows this radioisotope, ^{131}I, with a short half-life of 8.021 days.

9. 4 In Table N locate the nuclide ^{3}H. The decay mode is shown to be negative beta decay. Table O shows the notation and symbol for this particle.

10. 2 ^{90}Sr has a half-life of 29.1 y. Two half-lives would be 58.2 y. After two half-lives, 1/4 (25%) of the original radioactive material remains.

11. Answer: The half-life of cesium-137 is short, and the sample would almost be entirely decayed after 10,000 years. *or*   The half-life of strontium-90 is short, and the sample would almost be entirely decayed after 10,000 years.

Explanation: Cesium-137 has a half-life of 30.2 years and strontium-90 has a half-life of 29.1 years, as shown by Table N. After 10 half-lives very little radiation is emitted from these original radioactive sources. These elements have relatively short half-lives. Therefore, after 10 half-lives, about 300 years, the radiation would be very weak. After 10,000 years, just about all of these radioactive atoms would have decayed.

12. Answer: Krypton-85 decays by beta emission, while francium-220 decays by alpha emission.

Explanation: Table N shows that these two radioactive elements undergo radioactive decay by different decay modes.

13. *a)* Answer: natural transmutation *or* transmutation *or* beta decay *or* radioactive decay

Explanation: Table N shows that C-14 is a radioisotope that undergoes decay by beta emission. This radioactive decay process goes by the term of natural transmutation.

b) Answer: A C-13 atom has seven neutrons and a C-12 atom has six neutrons.
  or An atom of C-13 and an atom of C-12 have different numbers of neutrons.
  or The number of neutrons is different.

Explanation: To determine the number of neutrons in the nucleus of an atom, subtract the atomic number from the mass number.

c) Answer: 5715 y

Explanation: This is given in Table N under half-life.

14. Answer: 17,145 y

Explanation: If $\frac{1}{8}$ of the original material remains, it has undergone decay for 3 half-lives ($\frac{1}{2} \times \frac{1}{2} \times \frac{1}{2} = \frac{1}{8}$). Table N shows the half-life of C-14 is 5,715 years. Three half-lives is 17,145 y.

15. Answer: 6,396 y

Explanation: The mass of the original radioactive sample was 10.0 grams. After the 1st half-life, 5.0 g of the original radioactive sample would remain. After the 2nd half-life, 2.5 g would remain; after the 3rd half-life 1.25 g would remain; after the 4th half-life, 0.625 g would still be radioactive. Since half-life is constant, 4 half-lives of ^{226}Ra is 6,396 y (1,599 y × 4).

Name	Notation	Symbol
alpha particle	^4_2He or $^4_2\alpha$	α
beta particle	$^0_{-1}\text{e}$ or $^0_{-1}\beta$	β^-
gamma radiation	$^0_0\gamma$	γ
neutron	^1_0n	n
proton	^1_1H or ^1_1p	p
positron	$^0_{+1}\text{e}$ or $^0_{+1}\beta$	β^+

Overview:

Radioactive elements emit particles and/or energy from their nuclei. These nuclear particles or energy have different effects on the nuclei of these elements. This radioactive decay process can be represented by nuclear equations, showing the identity of the reactants, products and the radiation released. In these equations, both charge and mass must be conserved. To balance a nuclear equation, the sum of the atomic or charge numbers on each side must be equal, and the sum of the mass numbers on each side must be equal.

The Table:

This table gives the Name, Notation (used in writing nuclear equations) and Symbol of the common types of radiation.

Going down this chart in order:
- The alpha particle (α) is a helium nucleus and is therefore positively charged.
- The beta particle (β^-) is an ordinary electron and is therefore negatively charged.
- Gamma radiation (γ), is the emission of pure energy from the nucleus – it carries no charge or mass.
- The neutron (n) is a neutral particle of unit mass found in the nucleus of atoms.
- The proton (p) is a hydrogen nucleus, a positive particle of unit mass found in the nucleus of atoms.
- The positron (β^+) is a positive electron.

As shown in the Notation section, the mass number is the number at the upper left and the atomic number is the number at the lower left. When a particle is emitted, conservation of these numbers must take place with the reactants and products. If the emission of the particle affects the atomic number, the identity of the element changes. The specifics of these changes are given on the next page.

Additional Information:

Going down the chart in order:

- During alpha decay or emission, the atomic number decreases by 2 and the mass number decreases by 4. Example: $^{226}_{88}Ra \rightarrow \, ^{222}_{86}Rn + ^{4}_{2}He$

- During negative beta decay or emission, the atomic number increases by 1 and the mass number remains the same. Example: $^{235}_{92}U \rightarrow \, + ^{235}_{93}Np \, + \, ^{0}_{-1}e$

- During gamma emission, both the atomic number and mass number remain the same. Example: $^{239}_{92}U \rightarrow \, + ^{239}_{92}U \, + \, ^{0}_{0}\gamma$

- During neutron emission, the atomic number remains the same and the mass number decreases by 1. Example: $^{226}_{88}Ra \rightarrow \, ^{225}_{88}U + ^{1}_{0}n$

- During proton emission, both the atomic number and mass number decrease by 1. Example: $^{53}_{27}Co \rightarrow \, ^{52}_{26}Fe + ^{1}_{1}H$

- During positron emission, also known as positive beta emission, the atomic number decreases by 1 and the mass number remains the same. Example: $^{58}_{29}Cu \rightarrow \, ^{58}_{28}Ni + ^{0}_{+1}e$

- In natural transmutation, the identity of a nucleus or element changes due to a change in the number of protons (atomic number) in the nucleus. Example: $^{239}_{94}Pu \rightarrow \, ^{235}_{92}U + ^{4}_{2}He$

- Artificial transmutation occurs when a stable (nonradioactive) nucleus is bombarded with particles, causing it to become radioactive. Example: $^{9}_{4}Be + ^{4}_{2}He \rightarrow \, ^{12}_{6}C + ^{1}_{0}n$

- Nuclear fusion is the combining of lightweight nuclei to produce a heavier nucleus. This usually involves hydrogen nuclei combining to produce a helium nucleus. This is the source of solar energy. Example: $^{2}_{1}H + ^{2}_{1}H \rightarrow \, ^{4}_{2}He + $ energy

- Nuclear fission is the splitting of a heavier nucleus into lighter weight nuclei. U-235 and Pu-239 are the most common elements to under go fission. This is the source of the energy produced in nuclear reactors. Example: $^{235}_{92}U + ^{1}_{0}n \rightarrow \, ^{144}_{54}Xe + ^{90}_{38}Sr + 2^{1}_{0}n + $ energy

- Nuclear reactions release large amounts of energy due to the conversion of some mass into energy according to Einstein's equation, $E = mc^2$.

- Alpha radiation, being the largest particle, has the weakest penetrating power.

- Gamma radiation, being massless and neutral, has the greatest penetrating power.

Symbols Used in Nuclear Chemistry

1. Which particle has the *least* mass?

   (1) ^4_2He (3) ^1_0n

   (2) ^1_1H (4) $^0_{-1}e$ 1 _____

2. In the reaction $^{239}_{93}Np \rightarrow ^{239}_{94}Pu + X$, what does X represent?

   (1) a neutron
   (2) a proton
   (3) an alpha particle
   (4) a beta particle   2 _____

3. Positrons are spontaneously emitted from the nuclei of

   (1) potassium-37
   (2) radium-226
   (3) nitrogen-16
   (4) thorium-232   3 _____

4. Given the nuclear equation:
   $$^{19}_{10}Ne \rightarrow X + ^{19}_9F$$

   Which particle is represented by X?

   (1) alpha   (3) neutron
   (2) beta   (4) positron   4 _____

5. Which list of radioisotopes contains an alpha emitter, a beta emitter, and a positron emitter?

   (1) C-14, N-16, P-32
   (2) Cs-137, Fr-220, Tc-99
   (3) Kr-85, Ne-19, Rn-222
   (4) Pu-239, Th-232, U-238   5 _____

6. Which product of nuclear decay has mass but no charge?

   (1) alpha particles
   (2) neutrons
   (3) gamma rays
   (4) beta positrons   6 _____

7. Which type of radioactive emission has a positive charge and weak penetrating power?

   (1) alpha particle   (3) gamma ray
   (2) beta particle   (4) neutron   7 _____

8. Which of these types of radiation has the greatest penetrating power?

   (1) alpha   (3) gamma
   (2) beta   (4) positron   8 _____

9. Which reaction is an example of natural transmutation?

   (1) $^{239}_{94}Pu \rightarrow ^{235}_{92}U + ^4_2He$

   (2) $^{27}_{13}Al + ^4_2He \rightarrow ^{30}_{15}P + ^1_0n$

   (3) $^{238}_{92}U + ^1_0n \rightarrow ^{239}_{94}Pu + 2^0_{-1}e$

   (4) $^{239}_{94}Pu + ^1_0n \rightarrow ^{147}_{56}Ba + ^{90}_{38}Sr + 3^1_0n$

        9 _____

10. Given the balanced equation representing a nuclear reaction:

    $$^{235}_{92}U + ^1_0n \rightarrow ^{142}_{56}Ba + ^{91}_{36}Kr + 3X + energy$$

    Which particle is represented by X?

    (1) $^0_{-1}e$ (3) ^4_2He
    (2) ^1_1H (4) ^1_0n 10 _____

11. The change that is undergone by an atom of an element made radioactive by bombardment with high-energy protons is called

(1) natural transmutation
(2) artificial transmutation
(3) natural decay
(4) radioactive decay   11 _____

12. One benefit of nuclear fission reactions is

(1) nuclear reactor meltdowns
(2) storage of waste materials
(3) biological exposure
(4) production of energy   12 _____

13. In a nuclear fusion reaction, the mass of the products is

(1) less than the mass of the reactants because some of the mass has been converted to energy
(2) less than the mass of the reactants because some of the energy has been converted to mass
(3) more than the mass of the reactants because some of the mass has been converted to energy
(4) more than the mass of the reactants because some of the energy has been converted to mass   13 _____

14. Given the nuclear equation: $^{235}_{92}U + ^{1}_{0}n \rightarrow ^{142}_{56}Ba + ^{91}_{36}Kr + 3^{1}_{0}n + energy$

a) State the type of nuclear reaction represented by the equation. _____

b) The sum of the masses of the products is slightly less than the sum of the masses of the reactants. Explain this loss of mass.

_____

_____

c) This process releases greater energy than an ordinary chemical reaction does. Name another type of nuclear reaction that releases greater energy than an ordinary chemical reaction.

_____

15. Using Reference Table N, complete the equation below for the nuclear decay of $^{226}_{88}Ra$. Include both atomic number and mass number for each particle.

$^{226}_{88}Ra \rightarrow$ _____ + _____

Symbols Used in Nuclear Chemistry

16. A beta particle may be spontaneously emitted from

    (1) a ground-state electron
    (2) a stable nucleus
    (3) an excited electron
    (4) an unstable nucleus   16 _____

17. Given the nuclear equation:

    $$^{14}_{7}N + X \rightarrow\ ^{16}_{8}O +\ ^{2}_{1}H$$

    What is particle X?

    (1) an alpha particle
    (2) a beta particle
    (3) a deuteron
    (4) a triton   17 _____

18. In the reaction $^{9}_{4}Be + X \rightarrow\ ^{6}_{3}Li +\ ^{4}_{2}He$, the X represents

    (1) $^{0}_{+1}e$ (3) $^{0}_{-1}e$
    (2) $^{1}_{1}H$ (4) $^{1}_{0}n$ 18 _____

19. Given the nuclear reaction:

    $$^{32}_{16}S +\ ^{1}_{0}n \rightarrow\ ^{1}_{1}H + X$$

    What does X represent in this reaction?

    (1) $^{31}_{15}P$ (3) $^{31}_{16}S$
    (2) $^{32}_{15}P$ (4) $^{32}_{16}S$ 19 _____

20. Given the nuclear equation:

    $$^{253}_{99}Es + X \rightarrow\ ^{1}_{0}n +\ ^{256}_{101}Md$$

    Which particle is represented by X?

    (1) $^{4}_{2}He$ (3) $^{1}_{0}n$
    (2) $^{0}_{-1}e$ (4) $^{0}_{+1}e$ 20 _____

21. Given the equation: $^{239}_{93}Np \rightarrow\ ^{239}_{94}Pu + X$

    When the equation is balanced correctly, which particle is represented by X?

    (1) $^{0}_{1}e$ (3) $^{2}_{1}H$
    (2) $^{1}_{1}H$ (4) $^{1}_{0}n$ 21 _____

22. Which of these particles has the greatest mass?

    (1) alpha   (3) neutron
    (2) beta   (4) positron   22 _____

23. Types of nuclear reactions include fission, fusion, and

    (1) single replacement
    (2) neutralization
    (3) oxidation-reduction
    (4) transmutation   23 _____

24. Nuclear fusion differs from nuclear fission because nuclear fusion reactions

    (1) form heavier isotopes from lighter isotopes
    (2) form lighter isotopes from heavier isotopes
    (3) convert mass to energy
    (4) convert energy to mass   24 _____

25. Energy is released during the fission of Pu-239 atoms as a result of the

    (1) formation of covalent bonds
    (2) formation of ionic bonds
    (3) conversion of matter to energy
    (4) conversion of energy to matter   25 _____

26. Which equation is an example of artificial transmutation?

(1) $_4^9\text{Be} + _2^4\text{He} \rightarrow _6^{12}\text{C} + _0^1\text{n}$
(2) $\text{U} + 3\text{ F}_2 \rightarrow \text{UF}_6$
(3) $\text{Mg(OH)}_2 + 2\text{ HCl} \rightarrow 2\text{ H}_2\text{O} + \text{MgCl}_2$
(4) $\text{Ca} + 2\text{ H}_2\text{O} \rightarrow \text{Ca(OH)}_2 + \text{H}_2$

26 _____

27. Which equation represents a fusion reaction?

(1) $_1^2\text{H} + _1^2\text{H} \rightarrow _2^4\text{He}$
(2) $_6^{14}\text{C} \rightarrow _{-1}^0\text{e} + _7^{14}\text{N}$
(3) $_{92}^{238}\text{U} + _2^4\text{He} \rightarrow _{94}^{241}\text{Pu} + _0^1\text{n}$
(4) $_0^1\text{n} + _{13}^{27}\text{Al} \rightarrow _{11}^{24}\text{Na} + _2^4\text{He}$

27 _____

28. Which list of nuclear emissions is arranged in order from the least penetrating power to the greatest penetrating power?

(1) alpha particle, beta particle, gamma ray
(2) alpha particle, gamma ray, beta particle
(3) gamma ray, beta particle, alpha particle
(4) beta particle, alpha particle, gamma ray

28 _____

29. Complete the equation below for the radioactive decay of $_{55}^{137}\text{Cs}$. Include both atomic number and mass number for each particle.

$$_{55}^{137}\text{Cs} \rightarrow \underline{\quad _{-1}^0 e \quad} + \underline{\quad _{56}^{137} Be \quad}$$

30. Complete the nuclear equation below for the decay of K-42. Your response must include the atomic number, the mass number, and the symbol of the missing particle.

$$_{19}^{42}\text{K} \rightarrow _{-1}^0\text{e} + \underline{\hspace{3cm}}$$

31. Identify particle X in the equation $_1^2\text{H} + _1^3\text{H} \rightarrow _2^4\text{He} + X + \text{energy}$. Your response must include the symbol, the atomic number, and mass number of the particle.

_____

32. Complete the nuclear equation below for the decay of C-14. Your response must include the atomic number, the mass number, and the symbol of the missing particle.

$$_6^{14}\text{C} \rightarrow _{-1}^0\text{e} + \underline{\hspace{3cm}}$$

33. Nuclear equation: $_{92}^{235}\text{U} + _0^1\text{n} \rightarrow _{38}^{92}\text{Sr} + \underline{\hspace{2cm}} + 2_0^1\text{n} + \text{energy}$

Write an isotopic notation for the missing product in the nuclear equation.

_____

34. Which nuclear emission has no mass and no charge? _____

Symbols Used in Nuclear Chemistry

1. 4   Choice 4 is a negative beta particle which is an ordinary electron. The mass of an electron is negligible compared to a proton or a neutron.

2. 4   In a nuclear reaction, the sum of the atomic numbers on each side must be equal and the sum of the mass numbers on each side must be equal. Particle X therefore must have a charge number of -1 and a mass number of 0. From Table O, this must be a beta particle. Np, element 93, is a man-made radioactive element that decays to Pu, element 94, by releasing a beta particle.

3. 1   Table O shows that a positron, also known as a positive beta particle, has the symbol β^+. Open to Table N and locate K-37. Under Decay Mode, it shows that this element undergoes positron emission. During positron emission, the atomic number decreases by 1 and the mass number remains the same.

4. 4   To balance the nuclear equation, particle X must have a $+1$ charge and zero mass. Using the notation column in Table O, a positron has a $+1$ charge and 0 mass. During positron emission, the atomic number decreases by 1 and the mass number remains the same.

5. 3   Table O gives the particles and their notations referred to in this question. Using Table N, find the radioisotopes given in choice 3. The Decay Mode shows that choice 3 contains an alpha emitter, a beta emitter and a positron emitter.

6. 2   Looking at Table O, the neutron has the symbol $_0^1 n$, indicating it has an atomic mass of 1 and carries no charge.

7. 1   The first particle shown in Table O is the alpha particle. This particle is a helium nucleus, consisting of 2 protons and 2 neutrons. This makes this particle positive in charge. The alpha particle is the largest particle that is emitted in radioactive decay. Being the largest particle, it would have the weakest penetrating power, giving up more energy with each particle it collides with.

8. 3   In the notation column of Table O, it shows that gamma radiation has no charge or mass. A massless, neutral entity can easily pass through matter without interacting with it. This makes the gamma radiation the most penetrating of the given choices.

9. 1   In natural radioactivity, also called natural transmutation, a radioactive element will undergo a spontaneous decay process involving the nucleus. An example of this decay is radioactive Pu-239 changing to U-235 by alpha emission. Choices 2 and 3 show artificial transmutation (see explanation for answer no. 11). Choice 4 shows the fission of a Pu-239 nucleus.

10. 4 In a nuclear reaction, the sum of the atomic numbers and the sum of the mass numbers on each side must be equal. Particle 3X must have a charge number of 0 and a mass number of 1 (making a total mass number of 3(1) = 3). From Table O this particle is a neutron.

11. 2 In artificial transmutation, the nucleus of a non-radioactive element is bombarded with high energy particles, such as alpha particles, protons or neutrons. The end product of artificial transmutation is the formation of new radioactive elements.

12. 4 When nuclear fission occurs, a very large amount of energy is released. Under controlled conditions, this energy can safely be harnessed, changed to electrical energy and distributed to benefit mankind.

13. 1 In any nuclear reaction, the mass of the products is less than the mass of the reactants. This difference in mass is converted into a large amount of energy according to Einstein's equation, $E = mc^2$.

14. *a*) Answer: nuclear fission

Explanation: A neutron is captured by the nucleus of a uranium atom. This causes uranium to undergo nuclear fission, producing two lighter elements, while giving off subatomic particles and energy.

b) Answer: Mass has been converted into energy.

Explanation: In any nuclear reaction, the mass of the products is less than the mass of the reactants. This difference in mass is converted into a large amount of energy according to Einstein's equation, $E = mc^2$.

c) Answer: Nuclear fusion *or* natural transmutation *or* radioactive decay *or* nuclear decay

Explanation: In a nuclear fusion reaction, lighter nuclei combine or unite to form a heavier nucleus. As in any nuclear reaction, the mass of the products is less than the mass of the reactants. This difference in mass has been converted into energy.

15. Answer: $^{226}_{88}\text{Ra} \rightarrow \, ^{4}_{2}\text{He} + \, ^{222}_{86}\text{Rn}$ *or* $^{226}_{88}\text{Ra} \rightarrow \, ^{222}_{86}\text{Rn} + \, ^{4}_{2}\alpha$

Explanation: Locate Ra-226 in Table N. It shows that this element undergoes alpha emission. Table O shows the notation of an alpha particle. The above equations show that the total atomic numbers and mass numbers are conserved.

Prefix	Number of Carbon Atoms
meth-	1
eth-	2
prop-	3
but-	4
pent-	5
hex-	6
hept-	7
oct-	8
non-	9
dec-	10

Overview:

Organic chemistry is the study of carbon compounds. Carbon has the ability to not only form bonds with other elements but also bond indefinitely with other carbon atoms. Therefore, there are an enormous number of organic compounds. To help with the study of organic compounds, a systematic means of naming these compounds has been developed. One of these rules is the use of prefixes indicating the number of carbon atoms in the molecule.

The Table:

This table gives the prefixes used to indicate the number of carbon atoms in the longest continuous chain of carbon atoms in the molecule. The prefixes for 5 through 10 carbon atoms are Greek numerical prefixes for that number of carbon atoms.

Additional Information:

- Methane (CH_4) is also known as natural gas.
- Propane (C_3H_6) is used as a fuel for gas grills.
- Butane (C_4H_{10}) is used in lighters for gas grills.
- Octane (C_8H_{18}) is used in the anti-knock rating of gasoline.

— Set 1 —

1. Which element must be present in an organic compound?

   (1) hydrogen
   (2) oxygen
   (3) carbon
   (4) nitrogen

   1 _____

2. What is the total number of carbon atoms in a molecule of ethanoic acid?

   (1) 1   (3) 3
   (2) 2   (4) 4

   2 _____

3. A molecule of butane and a molecule of 2-butene both have the same total number of

   (1) carbon atoms
   (2) hydrogen atoms
   (3) single bonds
   (4) double bonds

   3 _____

4. A molecule of octane has how many more carbon atoms that a molecule of methane?

   (1) 1   (3) 5
   (2) 3   (4) 7

   4 _____

5.
```
     H  H  H  H  H  H  H  H
     |  |  |  |  |  |  |  |
  H─ C─ C─ C─ C─ C─ C─ C─ C─ H
     |  |  |  |  |  |  |  |
     H  H  H  H  H  H  H  H
```

   The above hydrocarbon molecule would have a prefix of

   (1) meth-   (3) oct-
   (2) but-   (4) non-

   5 _____

— Set 2 —

6. Which would be the proper prefix of the following hydrocarbon

   $$CH_3CH_2CHCH_2$$

   (1) eth-   (3) but-
   (2) prop-   (4) penti-

   6 _____

7. The more carbon atoms a compound has, the more isomers it will have. Which prefix would indicate the greatest number of isomers?

   (1) hept-   (3) meth-
   (2) but-   (4) dec-

   7 _____

8. A molecule of a compound contains a total of 10 hydrogen atoms and has the general formula C_nH_{2n+2}. Which prefix is used in the name of this compound?

   (1) but-   (3) oct-
   (2) dec-   (4) pent-

   8 _____

9. What is the correct formula for pentene?

   (1) C_4H_8 (3) C_7H_{14}
   (2) C_4H_{12} (4) C_5H_{10}

   9 _____

10. Which hydrocarbon is unsaturated, containing a triple bond and has 8 carbon atoms?

    (1) butane   (3) octyne
    (2) butyne   (4) decyne

    10 _____

1. 3   All organic compounds must contain the element carbon. Organic chemistry is the study of compounds containing carbon.

2. 2   In Table P, it shows that organic compounds with the prefix of "eth-" will contain 2 carbon atoms.

3. 1   Butane, an alkane, and 2-butene, an alkene, are different organic molecules, possessing different properties. However, both molecules will contain 4 carbon atoms as indicated by the prefix but-.

4. 4   Using the prefixes in Table P, methane contains one carbon atom and octane contains 8 carbon atoms. Thus octane has 7 more carbon atoms than methane.

5. 3   This hydrocarbon molecule has 8 carbon atoms. The prefix for 8 carbon atoms in Table P is oct-.

Name	General Formula	Examples	
		Name	Structural Formula
alkanes	C_nH_{2n+2}	ethane	H H | | H—C—C—H | | H H
alkenes	C_nH_{2n}	ethene	H H \ / C=C / \ H H
alkynes	C_nH_{2n-2}	ethyne	H—C≡C—H

Note: n = number of carbon atoms

Overview:

Organic compounds may be classified in homologous series. A homologous series is a group of compounds having related structures and properties. Each member of a homologous series differs from the preceding by a common increment – this being one carbon atom and two hydrogen atoms. Any homologous series can be represented by a general formula.

The Table:

This table gives the Name, General Formula and Examples of the Structural Formula of three homologous series of hydrocarbons. A hydrocarbon is a compound composed of only carbon and hydrogen.

Homologous Series:

The *alkanes* are saturated hydrocarbons, having only single bonds between adjacent carbon atoms. The name of each alkane ends in –ane. The prefix from Table P indicates the number of carbon atoms. Their chemical formula must match the General Formula given in the table.

The *alkenes* are unsaturated hydrocarbons, having one double bond between two adjacent carbon atoms. The name of each alkene ends in –ene. The prefix from Table P indicates the number of carbon atoms. Their chemical formula must match the General Formula given in the table.

The *alkynes* are unsaturated hydrocarbons, having one triple bond between two adjacent carbons atoms. The name of each alkyne ends in –yne. The prefix from Table P indicates the number of carbon atoms. Their chemical formula must match the General Formula given in the table.

In the General Formula, n represents the number of carbon atoms in the molecule.

General Formula Examples:

(1) A hydrocarbon from the alkane series contains 4 carbon atoms. What is its molecular formula?

Solution: $n = 4$ and the general formula for alkanes is C_nH_{2n+2}.

Substitution gives: $C_4H_{2(4)+2} = C_4H_{10}$ (butane)

(2) What homologous series does the hydrocarbon C_3H_4 belong to?

Solution: The number of carbon atoms is 3, thus $n = 3$.

Substituting in the general formula for the alkyne series: $C_3H_{2(3)-2} = C_3H_4$

Additional Information:

- A double bond consists of two shared pairs of electrons (4 electrons). A triple bond consists of three shared pairs of electrons (6 electrons).

- Isomers are compounds with the same molecular formula (number of atoms of each element) but different structural formulas. Structural formulas show the different arrangement of the atoms in the molecules of the isomers of a given compound. The greater the number of carbon atoms in the molecule, the greater the number of isomers. The first hydrocarbon to show isomerism is butane. The structural formulas for the two isomers of butane (C_4H_{10}) are:

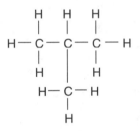

Normal butane or n-butane   **Isobutane or 2-methylpropane**

- Hydrocarbons contain covalent bonds and the molecules are nonpolar.

- Addition is the reaction in which one or more atoms are added to the site of a multiple bond in an unsaturated hydrocarbon, resulting in the formation of a saturated hydrocarbon.

- Substitution is the reaction in which one type of atom or group of atoms replaces another atom or group of atoms. This usually involves the replacement of one or more hydrogen atoms on a saturated hydrocarbon molecule.

- Polymerization is the chemical combination of a large number of molecules of a certain type, called monomers, to form a giant molecule, called a polymer. An example is $n\ C_2H_4 \rightarrow (C_2H_4)_n$, where the monomer is ethylene and the polymer is polyethylene

- As the molecular mass increases in a homologous series of hydrocarbons, the boiling point of the compounds increases.

1. Which element has atoms that can form single, double, and triple covalent bonds with other atoms of the same element?

    (1) hydrogen   (3) fluorine
    (2) oxygen   (4) carbon   1 _____

2. What is the general formula for the members of the alkane series?

    (1) C_nH_{2n} (3) C_nH_{2n-2}
    (2) C_nH_{2n+2} (4) C_nH_{2n-6} 2 _____

3. Which hydrocarbon is saturated?

    (1) propene   (3) butene
    (2) ethyne   (4) heptane   3 _____

4. Which formula represents a saturated hydrocarbon?

    (1) C_2H_2 (3) C_3H_4
    (2) C_2H_4 (4) C_3H_8 4 _____

5. A double carbon-carbon bond is found in a molecule of

    (1) pentane   (3) pentyne
    (2) pentene   (4) pentanol   5 _____

6. A straight-chain hydrocarbon that has only one double bond in each molecule has the general formula

    (1) C_nH_{2n-6} (3) C_nH_{2n}
    (2) C_nH_{2n-2} (4) C_nH_{2n+2} 6 _____

7. Which formula represents an unsaturated hydrocarbon?

    (1) C_2H_6 (3) C_5H_8
    (2) C_3H_8 (4) C_6H_{14} 7 _____

8. Which general formula represents the homologous series of hydrocarbons that includes the compound l-heptyne?

    (1) C_nH_{2n-6} (3) C_nH_{2n}
    (2) C_nH_{2n-2} (4) C_nH_{2n+2} 8 _____

9. A carbon-carbon triple bond is found in a molecule of

    (1) butane   (3) butene
    (2) butanone   (4) butyne   9 _____

10. Which formula represents propyne?

    (1) C_3H_4 (3) C_5H_8
    (2) C_3H_6 (4) C_5H_{10} 10 _____

11. What is the total number of electrons shared in the bonds between the two carbon atoms in a molecule of
    $$H - C \equiv C - H \ ?$$

    (1) 6   (3) 3
    (2) 2   (4) 8   11 _____

12. Which formula represents an alkene?

    (1) C_2H_6 (3) C_4H_{10}
    (2) C_3H_6 (4) C_5H_{12} 12 _____

13. Which compound is classified as a hydrocarbon?

    (1) ethane
    (2) ethanol
    (3) chloroethane
    (4) ethanoic acid   13 _____

14. Which formula represents an unsaturated hydrocarbon?

    (1)

    (2)

    (3)

    (4)   14 _____

15. Which compound is a saturated hydrocarbon?

    (1) hexane   (3) hexanol
    (2) hexene   (4) hexanal   15 _____

16. The isomers butane and methylpropane differ in their

    (1) total number of bonds per molecule
    (2) structural formulas
    (3) total number of atoms per molecule
    (4) molecular formulas   16 _____

17. Which formula is an isomer of butane?

    (1)

    (2)

    (3)

    (4)   17 _____

18. Given the structural formula:

    What is the IUPAC name of this compound?

    (1) propane   (3) propanone
    (2) propene   (4) propanal   18 _____

19. Which formula correctly represents the product of an addition reaction between ethene and chlorine?

    (1) CH_2Cl_2 (3) $C_2H_4Cl_2$
    (2) CH_3Cl (4) C_2H_3Cl 19 _____

Base your answers to question 20 using the information below and your knowledge of chemistry.

Ethene (common name ethylene) is a commercially important organic compound. Millions of tons of ethene are produced by the chemical industry each year. Ethene is used in the manufacture of synthetic fibers for carpeting and clothing, and it is widely used in making polyethylene. Low-density polyethylene can be stretched into a clear, thin film that is used for wrapping food products and consumer goods. High-density polyethylene is molded into bottles for milk and other liquids.

20. *a*) Explain, in terms of bonding, why ethene is an unsaturated hydrocarbon.

_____

b) Draw the structural formula for ethene.

21. Given the balanced equation for producing bromomethane: $Br_2 + CH_4 \rightarrow CH_3Br + HBr$

Identify the type of organic reaction shown. _____

22. Given the structural formula of pentane:

```
      H  H  H  H  H
      |  |  |  |  |
  H − C− C− C− C− C −H
      |  |  |  |  |
      H  H  H  H  H
```

In the space to the right, draw a structural formula for an isomer of pentane.

Base your answers to question 23 on the condensed structural formula below.

$$CH_3CH_2CHCH_2$$

23. In the space below, draw the structural formula for this compound.

Homologous Series of Hydrocarbons

24. Hydrocarbons are compounds that contain

    (1) carbon, only
    (2) carbon and hydrogen, only
    (3) carbon, hydrogen, and oxygen, only
    (4) carbon, hydrogen, oxygen,
        and nitrogen, only   24 ____

25. Which compound is an unsaturated
    hydrocarbon?

    (1) hexanal   (3) hexanoic acid
    (2) hexane   (4) hexyne   25 ____

26. Given the compound:

$$H - \underset{\underset{H}{|}}{C} = \underset{\underset{H}{|}}{C} - H$$

    The symbol = represents

    (1) one pair of shared electrons
    (2) two pairs of shared electrons
    (3) a single covalent bond
    (4) a coordinate covalent bond   26 ____

27. Which general formula represents
    the compound CH_3CH_2CCH?

    (1) C_nH_n (3) C_nH_{2n-2}
    (2) C_nH_{2n} (4) C_nH_{2n+2} 27 ____

28. Which formula represents an
    unsaturated hydrocarbon?

    (1) CH_2CHCl
    (2) CH_3CH_2Cl
    (3) $CH_3CH_2CH_3$
    (4) CH_3CHCH_2 28 ____

29. What is the correct name for the substance
    represented by the structural formula below?

    (1) acetylene   (3) ethene
    (2) benzene   (4) propene   29 ____

30. What is the maximum number of
    covalent bonds that can be formed
    by one carbon atom?

    (1) 1   (3) 3
    (2) 2   (4) 4   30 ____

31. Given the formula:

$$H-\underset{\underset{H}{|}}{C}-\underset{\underset{H}{|}}{C}=\underset{\underset{H}{|}}{C}-\underset{\underset{H}{|}}{C}-\underset{\underset{H}{|}}{C}-H$$

    What is the IUPAC name of
    this compound?

    (1) 2-pentene   (3) 2-butene
    (2) 2-pentyne   (4) 2-butyne   31 ____

32. Which compound is a saturated
    hydrocarbon?

    (1) CH_2CH_2 (3) CH_3CHO
    (2) CH_3CH_3 (4) CH_3CH_2OH 32 ____

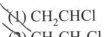

33. Which compound is an alkyne?

    (1) C_2H_2 (3) C_4H_8
    (2) C_2H_4 (4) C_4H_{10} 33 ____

34. Which formula represents an unsaturated hydrocarbon?

(1) C_5H_{12} (3) C_7H_{16}

(2) C_6H_{14} (4) C_8H_{14} 34 _____

35. In saturated hydrocarbons, carbon atoms are bonded to each other by

(1) single covalent bonds, only
(2) double covalent bonds, only
(3) alternating single and double covalent bonds
(4) alternating double and triple covalent bonds   35 _____

36. What is the IUPAC name of the organic compound that has the formula shown below?

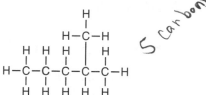

5 Carbons

(1) 1,1-dimethylbutane
(2) hexane
(3) 2-methylpentane
(4) 4-methylpentane   36 _____

37. Which structural formula correctly represents a hydrocarbon molecule?

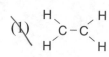
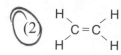

37 _____

4 bonds

38. Which structural formula represents an unsaturated hydrocarbon?

38 _____

39. What is the IUPAC name of the compound with the structural formula shown below?

(1) 2-pentene   (3) 2-pentyne
(2) 3-pentene   (4) 3-pentyne   39 _____

40. The three isomers of pentane have different

(1) formula masses
(2) molecular formulas
(3) empirical formulas
(4) structural formulas   40 _____

41. Which compound has an isomer?

41 _____

42. Which structural formula represents 2-pentyne?

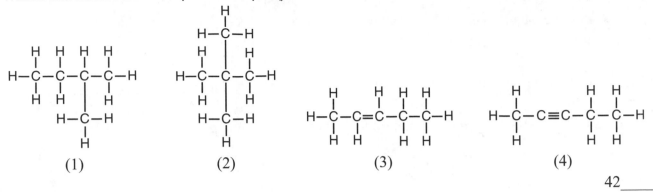

(1) (2) (3) (4)

42_____

43. Which Lewis electron-dot diagram represents chloroethene?

H H   ..
H:C::C:Cl:
(1)

H H ..
H:C:C:Cl:
.. ..
(2)

H H
.. .. ..
H:C::C:Cl:
H H
(3)

H
.. ..
H:C:Cl:
.. ..
H
(4)

43_____

44. The structural formula shown below represents butane. In the space provided draw an isomer of butane.

```
    H   H   H   H
    |   |   |   |
H — C — C — C — C — H
    |   |   |   |
    H   H   H   H
```

45. Give the formula for:

a) heptene

$C_7 H_{14}$

b) pentane

$C_5 H_{12}$

c) hexene

$C_6 H_{12}$

d) propyne

$C_3 H_4$

46. To which homologous series does $CH_3CH_2CH_2CH_3$ belong? _____alkane_____

$C_4 H_{10}$.

Base your answers to question 47 on the information below.

Given the balanced equation for an organic reaction between butane and chlorine that takes place at 300.°C and 101.3 kilopascals: $C_4H_{10} + Cl_2 \rightarrow C_4H_9Cl + HCl$

47. *a)* Identify the type of organic reaction shown. _____Substitution_____

b) In the space below, draw a structural formula for the organic product.

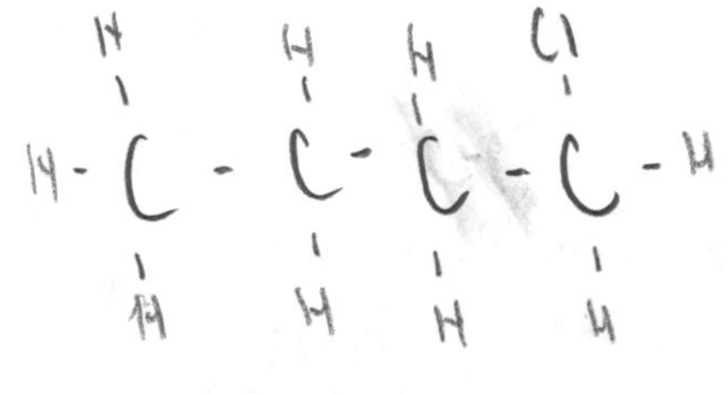

Base your answers to question 48 on the equation below, which represents an organic compound reacting with bromine.

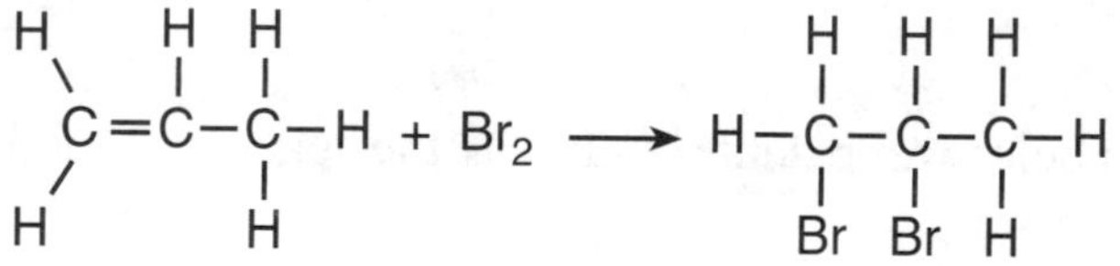

48. *a)* What is the IUPAC name for the organic compound that reacts with Br_2? __propyne__

b) What type of organic reaction is represented by this equation? __addition__

Base your answer to question 49 on the structural formula for octane shown to the right.

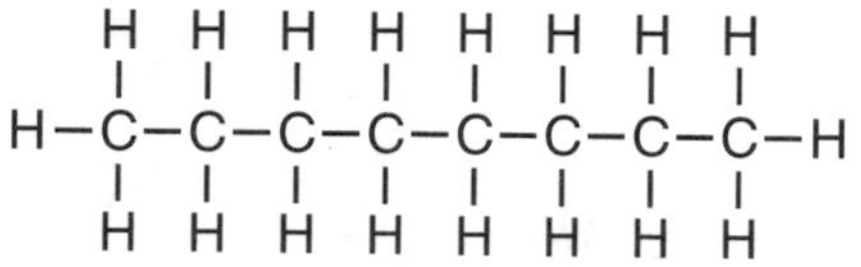

49. One isomer of octane is 2,2,4-trimethylpentane. In the space, draw a structural formula for 2,2,4-trimethylpentane.

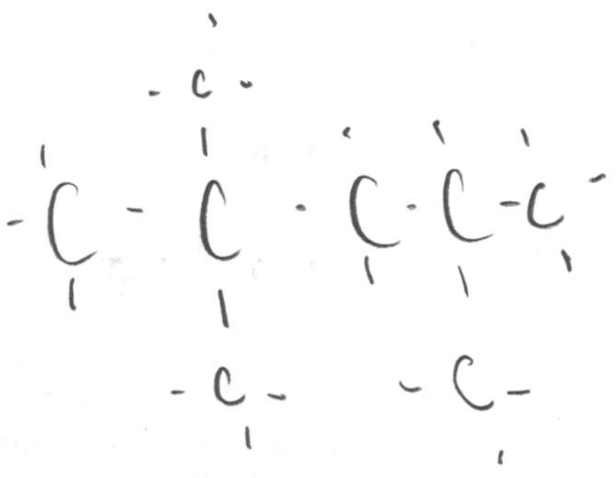

Set 1

1. 4   Hydrocarbons are organic compounds that contain only atoms of carbon and hydrogen. These organic compounds are then classified into different series based on their related structures and properties. Open to Table Q and here it shows the different structures and bonding capabilities of the carbon atom. The alkanes contain single bonds, the alkenes double bonds, and the alkynes triple bonds.

2. 2   As shown in Table Q, the general formula for the alkanes is C_nH_{2n+2} where n is the number of carbon atoms in the molecule.

3. 4   In saturated hydrocarbons there are only single bonds between adjacent carbon atoms and are members of the alkanes (see Table Q). Alkanes have the suffix of –ane. Heptane is an alkane and would contain 7 carbon atoms and 16 single bonded hydrogen atoms ($C_7H_{2(7)+2}$). The other given choices are members of the alkene or alkyne series.

4. 4   A saturate hydrocarbon must be a member of the alkane series. As shown in Table Q, the general formula for the alkanes is C_nH_{2n+2}. Choice 4, C_3H_8 fits this general formula for alkanes and is propane a saturated hydrocarbon.

5. 2   Open to Table Q and notice the given Structural Formulas. The alkenes (having a suffix of –ene) are shown as having a double bond between two adjacent carbon atoms.

6. 3   A hydrocarbon double bond is found in the alkenes. The general formula for alkenes is C_nH_{2n}.

7. 3   Unsaturated hydrocarbons contain either a double or triple bond, and are members of the alkene and alkyne series respectively. C_5H_8 is a member of the alkyne series, satisfying their general formula C_nH_{2n-2}. The other choices are all members of the alkanes, having a general formula of C_nH_{2n+2}.

8. 2   In Table Q it shows that a member of the alkyne series will have a triple bond between two carbons atoms and its name will have a suffix of -yne. A hydrocarbon molecule containing 7 carbon atoms will have the prefix of hept- (see Table P). Substituting into the general formula of the alkynes, C_nH_{2n-2} where $n = 7$, 1-heptyne will have the formula of C_7H_{12}.

9. 4   All alkynes contain a triple bond between two carbon atoms. This triple bond is diagrammed in the Structural Formula columnfor alkynes. All alkynes will have –yne as a suffix.

10. 1   Table P shows that an organic compound having 3 carbon atoms will have the prefix of prop-.
Table Q shows that a organic compound having the suffix of -yne belongs to the alkyne series
having a general formula of C_nH_{2n-2}. Substituting 3 for n into this general formula gives,
$C_3H_{2(3)-2}$ or C_3H_4, which is propyne.

11. 1   The given molecule is a member of the alkyne series due to the triple bond between the two
carbon atoms. Each covalent bond between the two carbon atoms consists of a pair of electrons.
Since this is a triple bond, a total of 6 electrons are being shared by the two carbon atoms. In
the diagram, each dash represents a covalent bond (a shared pair of electrons).

12. 2   The general formula for alkenes is C_nH_{2n}. Substituting 3 for n, the resulting alkene would be
C_3H_6, being propene. All other choices are alkanes.

13. 1   Hydrocarbons are organic compounds that contains only atoms of carbon and hydrogen. These
organic compounds are then classified into different groups based on their related structures and
properties. Ethane is a saturated hydrocarbon molecule of the alkane series having the formula
of C_2H_6.

14. 2   An unsaturated hydrocarbon will contain a double or triple bond between two carbon atoms.
Choice 2 is an unsaturated hydrocarbon due to its double bond and must be a member of the
alkenes. This molecule is ethene and is shown in Table Q. Choice 3 is an aldehyde and choice
4 is a halide or halocarbon, as shown in Table R.

15. 1   Saturated hydrocarbons are organic compounds that have single bonds between adjacent C
atoms. The alkane series are saturated hydrocarbons having the general formula C_nH_{2n+2} where
n is the number of carbon atoms. Hexane (C_6H_{14}) is a member of the alkane series.

16. 2   Isomers are compounds that have the same molecular formula, but show a different structural
formula. Referring to Table P and Q, butane is a member of the alkane series having the general
formula of C_nH_{2n+2}. Therefore, butane will have the molecular formula of C_4H_{10}. Since
methylpropane is an isomer of butane, it will have the same molecular formula (C_4H_{10}), but
a different structural formula.

17. 4   Isomers are compounds with the same molecular formula (number of atoms of each element)
but different structural formulas. Structural formulas show the different arrangement of the
atoms in the molecules of the isomers of a given compound. All isomers of butane, a member
of the alkane series, are saturated hydrocarbons (having single bonds between carbon atoms)
and will have 4 carbon atoms (see Table P).

18. 2   The structural formula shows a hydrocarbon with a double bond between two carbon atoms,
making it a member of the alkene series. The given molecular formula is C_3H_6. In Table P,
3 carbon atoms would be assigned the prefix of prop-. In Table Q, the alkenes are assigned the
suffix of –ene. This makes the IUPAC name for this compound propene.

19. 4   Ethene (C_2H_4) is a member of the alkene series. Each alkene contains a double bond between two carbon atoms, as shown in Table Q. When a chlorine atom is substituted in this molecule, a hydrogen atom gets replaced with a chlorine atom. The molecular formula now becomes C_2H_3Cl and is called chloroethene.

20. *a)*   Answer: has a carbon-carbon double bond   *or*   two carbons share four electrons   *or* $C = C$

Explanation: Ethene is a member of the alkene series. As shown in Table Q, the alkenes will always contain a double carbon to carbon bond. This bond makes this molecule a unsaturated hydrocarbon.

b)   Explanation: Ethene is a member of the alkenes. These molecules always contain a double carbon to carbon bond. With the prefix of eth-, this molecule will have 2 carbon atoms (see Table P). Using the general formula from Table Q for the alkenes (C_nH_{2n}) were *n* is 2, the molecule will contain 4 hydrogen atoms.

21. Acceptable responses include, but not limited to: Substitution   *or*   bromination   *or*   halogenation

Explanation: The molecular formula CH_4 represents a methane molecule, a saturated hydrocarbon (see Table P and Table Q – alkane series). A bromine atom has replaced a hydrogen atom in methane producing CH_3Br. This process is referred to as substitution, which is typical of saturated hydrocarbons. Because bromine is a halogen (Group 17 element), it may also be called halogenation and since bromine is doing the substitution, it may also be called bromination.

22.

Explanation: Isomers are compounds with the same molecular formula (number of atoms of each element) but different structural formulas. Structural formulas show the different arrangement of the atoms in the molecules of the isomers of a given compound. Notice both pentane and the isomer have a molecular formula of C_5H_{12}.

23.

Explanation: The molecular formula of this molecule is C_4H_8. This formula fits the general formula for members of the alkenes (C_nH_{2n}). All alkenes will contain a carbon-carbon double bond. A structural formula for this molecule must have these requirements.

Overview:

In Table Q you were introduced to homologous series of hydrocarbons. In this table other homologous series of organic compounds are shown. These organic compounds are formed when one or more hydrogen atoms of a hydrocarbon were replaced by other atoms or groups of atoms. These groups of atoms are called functional groups. A functional group is a particular arrangement of atoms, which gives characteristic properties to an organic molecule.

The Table:

The table shows the Class of Compound, Functional Group, General Formula and an Example of a compound containing that functional group. The R (in the General Formula) represents what is left of the hydrocarbon molecule after one or more of the hydrogen atoms of that molecule have been replaced by a different atom or group of atoms called the functional group. The name of the new compound is based on the nature of the functional group. The names of the compounds in each series are based on the name of the hydrocarbon containing the same number of carbon atoms as in the longest continuous chain of carbon atoms in the molecule that contains the functional group.

Class of Compound	Functional Group	General Formula	Example
halide (halocarbon)	—F (fluoro-) —Cl (chloro-) —Br (bromo-) —I (iodo-)	R—X (X represents any halogen)	$CH_3CHClCH_3$ 2-chloropropane
alcohol	—OH	R—OH	$CH_3CH_2CH_2OH$ 1-propanol
ether	—O—	R—O—R'	$CH_3OCH_2CH_3$ methyl ethyl ether
aldehyde	$-\overset{\overset{O}{\|\|}}{C}-H$	$R-\overset{\overset{O}{\|\|}}{C}-H$	$CH_3CH_2\overset{\overset{O}{\|\|}}{C}-H$ propanal
ketone	$-\overset{\overset{O}{\|\|}}{C}-$	$R-\overset{\overset{O}{\|\|}}{C}-R'$	$CH_3CCH_2CH_2CH_3$ 2-pentanone
organic acid	$-\overset{\overset{O}{\|\|}}{C}-OH$	$R-\overset{\overset{O}{\|\|}}{C}-OH$	$CH_3CH_2\overset{\overset{O}{\|\|}}{C}-OH$ propanoic acid
ester	$-\overset{\overset{O}{\|\|}}{C}-O-$	$R-\overset{\overset{O}{\|\|}}{C}-O-R'$	$CH_3CH_2COCH_3$ methyl propanoate
amine	$-\overset{\|}{N}-$	$R-\overset{\overset{R'}{\|}}{N}-R''$	$CH_3CH_2CH_2NH_2$ 1-propanamine
amide	$-\overset{\overset{O}{\|\|}}{C}-\overset{\|}{N}H$	$R-\overset{\overset{O}{\|\|}}{C}-\overset{\overset{R'}{\|}}{N}H$	$CH_3CH_2\overset{\overset{O}{\|\|}}{C}-NH_2$ propanamide

Note: R represents a bonded atom or group of atoms.

To better understand how this table works, the first two Classes of Compounds (halide and alcohol) will be explained in detail. (See next page.)

Halides:

The first row shows the class of compounds called the halides or halocarbons. A halide is produced when a Group 17 element (the halogens) replaces a hydrogen atom in a hydrocarbon molecule. These elements are shown as the functional group. The general formula for a halide is then R–X, where X represents any halogen. In the example shown, a chlorine atom has taken the place of one hydrogen atom from the hydrocarbon molecule. In naming the compound, the name of the halogen is modified to end in –o, hence chloro-. The rest of the name is derived from the name of the corresponding hydrocarbon with the same number of carbon atoms, in this case propane (3 carbon atoms, see Table P). The number 2 in the name represents the position of the chlorine atom in the chain of carbon atoms, in this case, the second from either end.

Alcohols:

The second row shows the class of compounds known as the alcohols. All alcohols must contain the –OH functional group. The general formula is then R–OH. The example shows this functional group (–OH) attached to the end of this molecule. The naming of the alcohols is done by replacing the final –e of the hydrocarbon name with –ol.

IUPAC Nomenclature:

As the table shows, each class of compounds has a different functional group. The name of the compound indicates the functional group present in the compound. The name of the parent hydrocarbon is usually modified by replacing the final –e of the name with a suffix indicating that class of compounds or functional group. *For example:* alcohols end in –ol   ketones end in –one   organic acids end in –oic

aldehydes end in –al   esters end in –ate   amines end in –ine

If necessary, the longest chain of carbon atoms is numbered to show the position of the functional group in that chain, using the smallest number.

Additional Information:

- The –OH of the alcohols does not ionize to produce the hydroxide ion (see Table E) that produces a basic solution.

- Esterification is the reaction between an acid and an alcohol to produce an ester and water. Organic esters are characterized by a pleasant taste and fragrance. They are used in artificial flavorings.

- Examples of organic compounds found in Table R:

    Ethyl alcohol (CH_3CH_2OH) is the alcohol found in alcoholic beverages and is the product of the fermentation reaction.

    2-propanol ($CH_3CHOHCH_3$) is rubbing alcohol.

    Diethyl ether ($C_2H_5OC_2H_5$) is operating room ether.

    Methanal (HCHO), an aldehyde, also called formaldehyde, is a liquid used to preserve animal specimens.

    Propanone (CH_3COCH_3), a ketone, also called acetone, is a common industrial solvent.

    Ethanoic acid (CH_3COOH), an organic acid, also called acetic acid, is in vinegar.

1. The functional group — COOH is found in

   (1) esters   (3) alcohols

   (2) aldehydes   (4) organic acids   1 _____

2. One molecule of propanol contains a total of

   (1) one –OH group

   (2) two –CH_3 groups

   (3) three –OH groups

   (4) three –CH_3 groups   2 _____

3. If a compound contains only one –OH functional group attached to the end carbon in the chain, it is classified as a

   (1) halide   (3) ether

   (2) alcohol   (4) ketone   3 _____

4. Given the structural formula:

   ```
     H  H     H  H
     |  |     |  |
   H–C––C––O––C––C–H
     |  |     |  |
     H  H     H  H
   ```

   The compound represented by this formula can be classified as an

   (1) organic acid   (3) ester

   (2) ether   (4) aldehyde   4 _____

5. Ethanol and dimethyl ether have different chemical and physical properties because they have different

   (1) functional groups

   (2) molecular masses

   (3) numbers of covalent bonds

   (4) percent compositions by mass   5 _____

6. Which compound an ketone?

   (1) ethyne   (3) propanone

   (2) methanal   (4) chloropropane   6 _____

7. The organic compound represented by the condensed structural formula

   $$CH_3CH_2CH_2CHO$$

   is classified as an

   (1) alcohol   (3) ester

   (2) aldehyde   (4) ether   7 _____

8. What is the IUPAC name for the compound that has the condensed structural formula $CH_3CH_2CH_2CHO$?

   (1) butanal   (3) propanal

   (2) butanol   (4) propanol   8 _____

9. The compounds 2-butanol and 2-butene both contain

   (1) double bonds, only

   (2) single bonds, only

   (3) carbon atoms

   (4) oxygen atoms   9 _____

10. Which of these compounds has chemical properties most similar to the chemical properties of ethanoic acid?

    (1) C_3H_7COOH

    (2) C_2H_5OH

    (3) $C_2H_5COOC_2H_5$

    (4) $C_2H_5OC_2H_5$ 10 _____

11. Given the formulas of four organic compounds:

(a)
```
    H  H  O
    |  |  ||
H−C−C−C−H
    |  |
    H  H
```

(c)
```
    H  O  H
    |  ||  |
H−C−C−C−H
    |     |
    H     H
```

(b)
```
    H  H  O
    |  |  ||
H−C−C−C−OH
    |  |
    H  H
```

(d)
```
    H  H  H
    |  |  |
H−C−C−C−H
    |  |  |
    H OH H
```

Which pair below contains an alcohol and an acid?

(1) *a* and *b* (3) *b* and *d*
(2) *a* and *c* (4) *c* and *d*   11 ____

12. Given the formula:
```
    H  H  H  O
    |  |  |  ||      H
H−C−C−C−C−N
    |  |  |      `H
    H  H  H
```

This compound is classified as

(1) an aldehyde
(2) an amide
(3) an amine
(4) a ketone   12 ____

Given the three organic structural formulas shown below:

```
    H  O  H
    |  ||  |
H−C−C−C−H
    |     |
    H     H
```

```
    H  O
    |  ||
H−C−C−OH
    |
    H
```

```
    H  OH H
    |  |  |
H−C−C−C−H
    |  |  |
    H  H  H
```

13. Which organic-compound classes are represented by these structural formulas, as shown from left to right?
(1) ester, organic acid, ketone   (3) ketone, aldehyde, alcohol
(2) ester, aldehyde, organic acid   (4) ketone, organic acid, alcohol   13____

Base your answers to question 14 using the information below and your knowledge of chemistry.

```
    H  O                              H  O     H  H  H
    |  ||                            |  ||    |  |  |
H−C−C−OH  +  X  →(catalyst)  H−C−C−O−C−C−C−H  +  H₂O
    |                                |        |  |  |
    H                                H        H  H  H
```

The incomplete equation above represents an esterification reaction. The alcohol reactant is represented by **X**.

14. *a*) On the accompanying structural formula, circle the acid functional group, only.

b) Write an IUPAC name for the reactant represented by its structural formula in this equation. _____

c) In the space in below, draw the structural formula for the alcohol represented by **X**.

15. Which compound is an alcohol?

   (1) propanal   (3) butane
   (2) ethyne   (4) methanol   15 ____

16. Which structural formula represents an alcohol?

   (1) 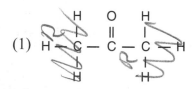 H — C — C — C — H

   (2) H — C — C — C — OH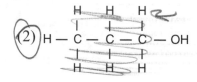

   (3) H — C — C
       | \\
       H   O, H

   (4) H — C — C
       | \\
       H   O, OH
       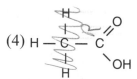
                                          16 ____

17. From the structural formulas in question 16, which choice contains the functional group of an organic acid?

   (1) 1   (3) 3
   (2) 2   (4) 4   17 ____

18. Which formula does not show a functional group?

   (1) $CH_3CH_2CH_2CHO$
   (2) $CH_3CH_2CH_2CH_3$
   (3) $CH_3CH_2CH_2COOH$
   (4) $CH_3CH_2COOCH_3$ 18 ____

19. Which structural formula represents an isomer of 1-propanol?

   (1) H — C — C — C — H

   (2) H — C — C — C
       with =O and H

   (3) H — C — C — C — H
       with OH on middle carbon

   (4) H — C — C — C
       with =O and OH   19 ____

20. Which functional group, when attached to a chain of carbon atoms, will produce an organic molecule with the characteristic properties of an aldehyde?

   (1) —C—OH   (3) —C—
       ‖   ‖
       O   O

   (2) —C—H   (4) —OH
       ‖
       O   20 ____

21. What is the total number of pairs of electrons shared between the carbon atom and the oxygen atom in a molecule of methanal?

   (1) 1   (3) 3
   (2) 2   (4) 4   21 ____

22. Which structural formula represents an ether?

(1)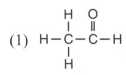

(2) 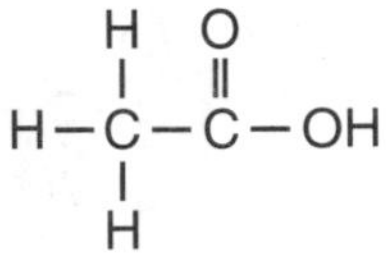 H–C–C–OH (with H and O, H below)

(3) 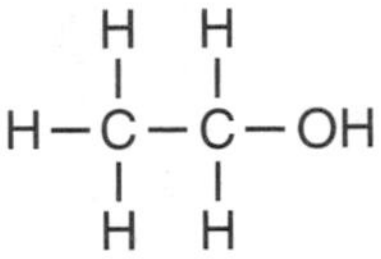 H–C–C–OH (with H H top, H H bottom)

(4) 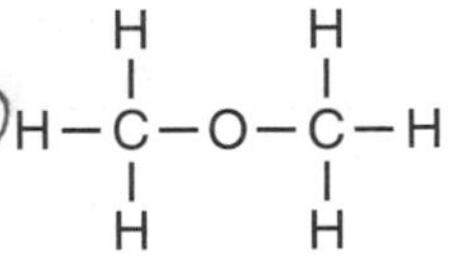 H–C–O–C–H (with H top both, H bottom both) — circled

22 ____

23. Which formula represents a ketone?

(1) HCOOH   (3) CH₃COCH₃

(2) HCHO   (4) CH₃CH₂OH

CH_3COCH_3 circled handwritten notes

23 ____

24. Which organic compounds are often used to create fragrances for the perfume industry?

(1) ethers   (3) alkanes — circled

(2) esters   (4) alkynes

24 ____

25. The reaction between an organic acid and an alcohol produces

(1) an aldehyde   (3) an ether

(2) a ketone   (4) an ester

25 ____

26. Which of these compounds has chemical properties most similar to the chemical properties of ethanoic acid?

(1) C_3H_7COOH (3) $C_2H_5COOC_2H_5$ — (1) circled

(2) C_2H_5OH (4) $C_2H_5OC_2H_5$

26 ____

27. Which formula represents an ether?

(1) 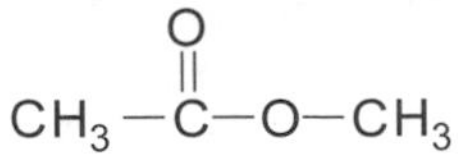 $CH_3-\overset{O}{\overset{\|}{C}}-O-CH_3$

(2) 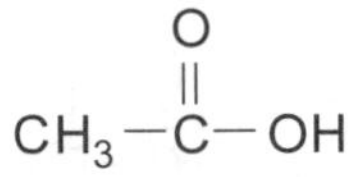 $CH_3-\overset{O}{\overset{\|}{C}}-OH$

(3) CH_3-O-CH_3 — circled

(4) CH_3-OH

27 ____

28. Given the structural formulas for two organic compounds:

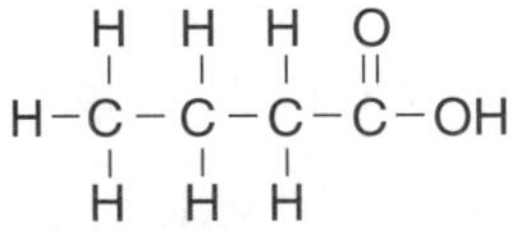

H–C–C–C–C–OH and H–C–C–C–O–C–H

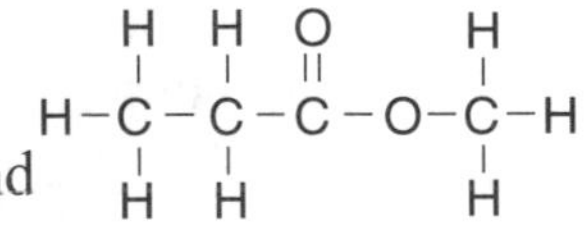

The differences in their physical and chemical properties are primarily due to their different

(1) number of carbon atoms
(2) number of hydrogen atoms
(3) molecular masses
(4) functional groups — circled

28 ____

29. Butanal and butanone have different chemical and physical properties primarily because of differences in their

(1) functional groups
(2) molecular masses
(3) molecular formulas
(4) number of carbon atoms per molecule

29 ____

30. Which structural formula is correct for 2-methyl-3-pentanol?

(1)   (2)   (3)   (4)

30 _____

31. Given the equation: butanoic acid + 1-pentanol $\xrightarrow{\text{catalyst}}$ water + X

    To which class of organic compounds does product X belong? _____ ester _____

    Base your answers to question 32 using the information below and your knowledge of chemistry.

                    Diethyl ether is widely used as a solvent.

32. a) In the space provided below, draw the structural formula for diethyl ether.

    b) In the space provided below, draw the structural formula for an alcohol that is an isomer of diethyl ether.

Set 1

1. **4** Organic acids contain the functional group –COOH. In Table R, find organic acid in the Class of Compound column. It is the functional group that gives the properties to the organic acid molecule.

2. **1** Open to Table R. In the Class of Compound column, find the alcohol row. Alcohols contain the –OH functional group. In the Example column, it shows that alcohol names have the suffix –ol. Thus, propanol must be an alcohol. Also, in the Example column, the formula for 1-propanol is given.

3. **2** As shown in Table R, the functional group of alcohols is –OH. It is this functional group that accounts for the properties that are common to alcohols.

4. **2** In Table R in the Class of Compound column, go down to ether. Here it shows that the functional group for ethers is –O–, with a General Formula of R–O–R', where R represents an alkyl group, which is derived from a hydrocarbon. The given structural formula shows that it is an ether, where R is C_2H_5–.

5. **1** A functional group is a particular arrangement of atoms, which gives characteristic properties to an organic molecule. Thus, molecules that have different functional groups will have different properties.

6. **3** Ketone compounds end in –one. Open to Table R and locate the ketone row. Notice that in the given Example, the shown organic compound has the suffix of –one.

7. **2** In the Class of Compound column, locate the aldehyde row. These compounds always contain the functional group of –CHO. This functional group is found in the given condensed structural formula, making it an aldehyde.

8. **1** The compound has a functional group of –CHO, making it a member of the aldehydes, as shown in Table R. The formula shows 4 carbon atoms giving it an organic prefix of but- (see Table P). The corresponding 4 carbon hydrocarbon would be butane (see Table Q). The final –e is dropped and replaced with -al to to name an aldehyde. Therefore, the name of the compound is butanal.

9. **3** Butene is an unsaturated alkene that has a double bond and contains only hydrogen and carbon atoms. Butanol is an alcohol with the functional group of –OH (having single bonds only) and will consists of carbon, hydrogen and oxygen atoms. Thus both molecules will have carbon atoms in common.

10. 1   Another organic acid would have similar chemical properties as ethanoic acid. Table R indicates that all organic acids contain the –COOH functional group. Answer 1 contains this functional group and is butanoic acid.

11. 3   The functional group of an alcohol is –OH. Choice (d) then must be an alcohol. Table R shows that an organic acid has the functional group –COOH. Choice (b) then must be an organic acid.

12. 2   The given functional group for an amide is –CONH, matches the one shown in the given formula. This compound is butanamide.

13. 4   The first structural formula shows the functional group of a ketone, the second formula shows the functional group for an organic acid, and the third formula shows the functional group for an alcohol.

14. *a*) Answer:

Explanation: The functional group for organic acids is –COOH as shown in Table R, organic acid row.

b) Answer: ethanoic acid   *or*   acetic acid

Explanation: Esterification is the reaction between an acid and an alcohol to produce and ester and water. The first reactant shows the functional group for an organic acid, –COOH.

c)

Explanation: The reactant **X** must contain 3 carbon atoms to maintain the same number of carbon atoms as the product side. Since **X** is an alcohol, the functional group of –OH must be attached to the molecule.

Periodic Table
of
The Elements

Periodic Table of the Elements

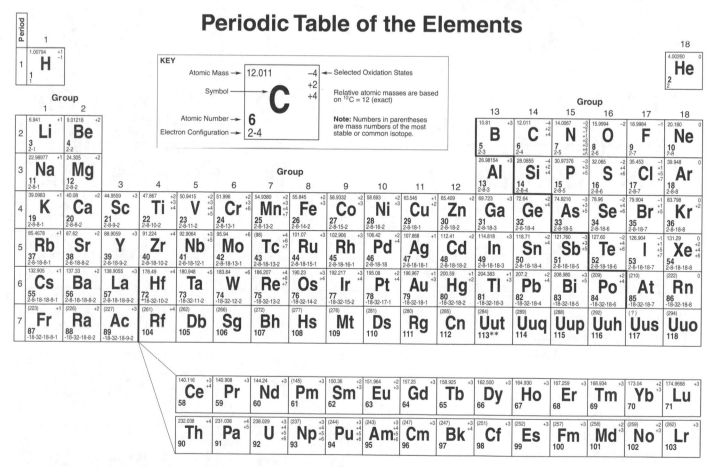

*denotes the presence of (2-8-) for elements 72 and above

**The systematic names and symbols for elements of atomic numbers 113 and above
will be used until the approval of trivial names by IUPAC.

Periodic Table of The Elements

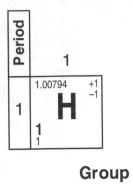

Period

1

| 1 | 1.00794 +1 −1 **H** 1 1 |

KEY

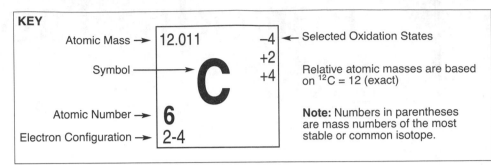

Atomic Mass → 12.011 −4 ← Selected Oxidation States
+2
Symbol → **C** +4

Relative atomic masses are based on $^{12}C = 12$ (exact)

Atomic Number → 6
Electron Configuration → 2-4

Note: Numbers in parentheses are mass numbers of the most stable or common isotope.

Group

	1	2	3	4	5	6	7	8	9
2	6.941 +1 **Li** 3 2-1	9.01218 +2 **Be** 4 2-2							
3	22.98977 +1 **Na** 11 2-8-1	24.305 +2 **Mg** 12 2-8-2					**Group**		
4	39.0983 +1 **K** 19 2-8-8-1	40.08 +2 **Ca** 20 2-8-8-2	44.9559 +3 **Sc** 21 2-8-9-2	47.867 +2 +3 +4 **Ti** 22 2-8-10-2	50.9415 +2 +3 +4 +5 **V** 23 2-8-11-2	51.996 +2 +3 +6 **Cr** 24 2-8-13-1	54.9380 +2 +3 +4 +7 **Mn** 25 2-8-13-2	55.845 +2 +3 **Fe** 26 2-8-14-2	58.9332 +2 +3 **Co** 27 2-8-15-2
5	85.4678 +1 **Rb** 37 2-8-18-8-1	87.62 +2 **Sr** 38 2-8-18-8-2	88.9059 +3 **Y** 39 2-8-18-9-2	91.224 +4 **Zr** 40 2-8-18-10-2	92.9064 +3 +5 **Nb** 41 2-8-18-12-1	95.94 +6 **Mo** 42 2-8-18-13-1	(98) +4 +6 +7 **Tc** 43 2-8-18-13-2	101.07 +3 **Ru** 44 2-8-18-15-1	102.906 +3 **Rh** 45 2-8-18-16-1
6	132.905 +1 **Cs** 55 2-8-18-18-8-1	137.33 +2 **Ba** 56 2-8-18-18-8-2	138.9055 +3 **La** 57 2-8-18-18-9-2	178.49 +4 **Hf** 72 *18-32-10-2	180.948 +5 **Ta** 73 -18-32-11-2	183.84 +6 **W** 74 -18-32-12-2	186.207 +4 +6 +7 **Re** 75 -18-32-13-2	190.23 +3 +4 **Os** 76 -18-32-14-2	192.217 +3 +4 **Ir** 77 -18-32-15-2
7	(223) +1 **Fr** 87 -18-32-18-8-1	(226) +2 **Ra** 88 -18-32-18-8-2	(227) +3 **Ac** 89 -18-32-18-9-2	(261) +4 **Rf** 104	(262) **Db** 105	(266) **Sg** 106	(272) **Bh** 107	(277) **Hs** 108	(276) **Mt** 109

* denotes the presence of (2-8-) for elements 72 and above

** The systematic names and symbols for elements of atomic numbers 113 and above will be used until the approved of trivial names by IU PAC.

140.116 +3 +4 **Ce** 58	140.908 +3 **Pr** 59	144.24 +3 **Nd** 60	(145) +3 **Pm** 61	150.36 +2 +3 **Sm** 62
232.038 +4 **Th** 90	231.036 +4 +5 **Pa** 91	238.029 +3 +4 +5 +6 **U** 92	(237) +3 +4 +5 +6 **Np** 93	(244) +3 +4 +5 +6 **Pu** 94

Periodic Table of The Elements

Group

18
4.00260   0
He
2
2

	13	14	15	16	17	18
	10.81 +3	12.011 −4 +2 +4	14.0067 −3 −2 −1 +2 +3 +4 +5	15.9994 −2	18.9984 −1	20.180 0
	B	**C**	**N**	**O**	**F**	**Ne**
	5 2-3	6 2-4	7 2-5	8 2-6	9 2-7	10 2-8
	26.98154 +3	28.0855 −4 +2 +4	30.97376 −3 +3 +5	32.065 −2 +3 +4 +6	35.453 −1 +1 +5 +7	39.948 0
	Al	**Si**	**P**	**S**	**Cl**	**Ar**
	13 2-8-3	14 2-8-4	15 2-8-5	16 2-8-6	17 2-8-7	18 2-8-8

10	11	12	13	14	15	16	17	18
58.693 +2 +3	63.546 +1 +2	65.409 +2	69.723 +3	72.64 +2 +4	74.9216 −3 +3 +5	78.96 −2 +4 +6	79.904 −1 +1 +5	83.798 0 +2
Ni	**Cu**	**Zn**	**Ga**	**Ge**	**As**	**Se**	**Br**	**Kr**
28 2-8-16-2	29 2-8-18-1	30 2-8-18-2	31 2-8-18-3	32 2-8-18-4	33 2-8-18-5	34 2-8-18-6	35 2-8-18-7	36 2-8-18-8
106.42 +2 +4	107.868 +1	112.41 +2	114.818 +3	118.71 +2 +4	121.760 −3 +3 +5	127.60 −2 +4 +6	126.904 −1 +1 +5 +7	131.29 0 +2 +4 +6
Pd	**Ag**	**Cd**	**In**	**Sn**	**Sb**	**Te**	**I**	**Xe**
46 2-8-18-18	47 2-8-18-18-1	48 2-8-18-18-2	49 2-8-18-18-3	50 2-8-18-18-4	51 2-8-18-18-5	52 2-8-18-18-6	53 2-8-18-18-7	54 2-8-18-18-8
195.08 +2 +4	196.967 +1 +3	200.59 +1 +2	204.383 +1 +3	207.2 +2 +4	208.980 +3 +5	(209) +2 +4	(210)	(222) 0
Pt	**Au**	**Hg**	**Tl**	**Pb**	**Bi**	**Po**	**At**	**Rn**
78 -18-32-17-1	79 -18-32-18-1	80 -18-32-18-2	81 -18-32-18-3	82 -18-32-18-4	83 -18-32-18-5	84 -18-32-18-6	85 -18-32-18-7	86 -18-32-18-8
(281)	(280)	(285)	(284)	(289)	(288)	(292)	(?)	(294)
Ds	**Rg**	**Cn**	**Uut**	**Uuq**	**Uup**	**Uuh**	**Uus**	**Uuo**
110	111	112	113**	114	115	116	117	118

151.964 +2 +3	157.25 +3	158.925 +3	162.500 +3	164.930 +3	167.259 +3	168.934 +3	173.04 +2 +3	174.9668 +3
Eu	**Gd**	**Tb**	**Dy**	**Ho**	**Er**	**Tm**	**Yb**	**Lu**
63	64	65	66	67	68	69	70	71
(243) +3 +4 +5 +6	(247) +3	(247) +3 +4	(251) +3	(252) +3	(257) +3	(258) +2 +3	(259) +2 +3	(262) +3
Am	**Cm**	**Bk**	**Cf**	**Es**	**Fm**	**Md**	**No**	**Lr**
95	96	97	98	99	100	101	102	103

Overview:

The Periodic Table is a systematic arrangement or classification of the elements. In the late 1800's, scientists noticed regularities in the properties of the elements, prompting them to arrange the elements based upon these recurring properties. It has passed through many stages of development in reaching its present form. This arrangement enabled scientists to predict the chemical and physical properties of elements that had not yet been discovered. This table will enable you to determine the identity of an element when given the properties of that element. The modern Periodic Law states that the properties of the elements are periodic functions of the atomic number or nuclear charge.

The Key:

Symbol – The symbol is a shorthand method of indicating the identity of the element. It consists of one or two letters from its name in English, or in many cases, from its name in Latin. The first letter is always capitalized, and the second letter (if present) is always written in lower case.

Atomic mass – The atomic mass of an element is the weighted average mass of the naturally occurring isotopes of that element. It is weighted according to the percent occurrence of each isotope. Atomic mass is expressed in atomic mass units (u). The atomic mass unit is defined relative to the mass of a C-12 atom, assigned an atomic mass of exactly 12.000 u. One atomic mass unit is 1/12 the mass of the standard C-12 atom. Since most elements occur as a mixture of isotopes, their atomic masses are decimal or fractional. It is calculated by taking the sum of the products of the percent occurrence and actual mass of each isotope. For example, Cl exists naturally as 75.40% Cl-35 (actual mass = 34.97) and 24.60% Cl-37 (actual mass = 36.97). The atomic mass of chlorine is: $(0.7540 \times 34.97) + (0.2460 \times 36.97) = 35.46$ u.

Atomic number – The atomic number of an element is the number of protons in the nucleus of an atom of that element. It is the property that identifies the element. Isotopes are atoms with the same atomic number but different mass numbers (the number of protons and neutrons in the nucleus). For example, C-12 and C-14 are isotopes. Both have 6 protons but C-12 has 6 neutrons while C-14 has 8 neutrons. The positive charge of a nucleus is equal to its atomic number.

Electron Configuration – The electron configuration given for the element shows the number of electrons in each principal energy level of an atom of that element in the ground state. The ground state is the state of lowest energy for the electrons in an atom.

Selected Oxidation States – The selected oxidation states (also referred to as oxidation number) show the charge on an atom of that element after gaining or losing an electron(s), or the apparent charge resulting from an unequal sharing of electrons with another atom. A loss of electrons (oxidation) results in a positive oxidation state. A gain of electrons (reduction) results in a negative oxidation state. Sharing of electrons with an atom of higher electronegativity results in a positive oxidation state. Sharing of electrons with an atom of lower electronegativity results in a negative oxidation state.

The Periodic Table:

Periods – The horizontal rows on the table are called periods. The properties of the elements change systematically through a period. With increasing atomic number in a given period, the properties of the elements change from metallic to metalloid to nonmetallic to inert (noble) gas. The exception is Period 1, where both elements are nonmetals. The period number is equal to the number of occupied energy levels of an atom in its ground state.

Groups – The vertical columns on the table are called groups or families. The elements in a group exhibit similar or related properties because they contain the same number of valence electrons. For example, Na and K would have similar chemical properties since they are both located in Group 1 and contain one valence electron. The Group 1 elements are called the alkali metals. Those in Group 2 are the alkaline earth elements. Group 17 elements are the halogens and those in Group 18 are the inert, rare or noble gases. The inert gases are chemically inactive, forming compounds only with the most active elements (F and O). All are monatomic gases at room conditions.

Valence Electrons – The period number is the same as the number of occupied energy levels of an atom in its ground state. The outer most energy level (last number in the electron configuration) is the valence level and the electrons in this energy level are the valence electrons. These electrons are the electrons lost, gained or shared during a chemical reaction, and therefore determine the chemical properties of the elements. For example: Lithium (Li) is in Period 2. It contains two principal energy levels and has one valence electron.

Classification of the Elements – The dark zig-zag line on the Periodic Table separates the metals from the nonmetals. Elements bordering this line, especially those to the right of the line, are called metalloids, and have properties of both metals and nonmetals. Most of the elements on the Periodic Table are metals and are found to the left of the zig-zag line. Metals are malleable, ductile, possess luster, and are good conductors of heat and electricity. All are solids at room temperature except mercury, which is a liquid. Metals lose electrons to form positive ions when reacting with nonmetals. The remaining elements are nonmetals and are found to the right of the zig-zag line. Nonmetals are brittle, lack luster and are nonconductors of heat and electricity. At room temperature, they exhibit all three phases of matter. Bromine is the only nonmetal that is a liquid at room temperature. Nonmetals tend to gain electrons during ionic bond formation, but they can form either positive or negative oxidation states during covalent bond formation with another nonmetal.

Additional Information:

*   Rounding off the atomic mass to the nearest whole number gives the atomic mass number or simply the mass number of the element. The mass number is the number of protons and neutrons (nucleons) in the nucleus of an atom of that element. Due to the weighting used in calculating the atomic mass of an element, this gives the mass number of the most common isotope of that element.

- To determine the number of neutrons in the nucleus of an atom, subtract the atomic number from the mass number.

- Any energy level higher than the ground state is referred to as an excited state. For example, the ground state configuration for calcium (Ca) is 2-8-8-2. A configuration of 2-8-7-3 would be an excited state since an electron from the 3rd energy level has been raised or excited to the 4th energy level.

- When an excited atom returns to the ground state, energy is released in the form of quanta, forming a bright line spectrum.

- With increasing atomic number within a given group, the elements become more metallic in properties.

- The selected oxidation states or numbers are used to determine the correct formula for a compound. In all formulas of compounds, the sum of the oxidation states must equal zero.

- The Lewis electron-dot diagram consists of the symbol of the element surrounded by dots or x's representing valence electrons only. Usually a maximum of two dots are positioned on each side of the symbol.

  The Lewis electron-dot diagram may be used to show the structure of molecules. The valence shell electrons are represented by dots or x's surrounding the symbol. The bonding electrons are shown between the atoms sharing the electrons. For single bonds, two electrons are shown, for double bonds, four electrons are shown and for triple bonds, six electrons are shown. Shared electron pairs may be represented by dashes between the symbols.

  Lewis electron-dot diagrams may also be used to represent ionic bond formation, is which the negative ion is placed in parentheses to indicate a transfer of electrons.

- Elements with atomic numbers greater than 109 have symbols consisting of 3 letters, each the first letter of the Greek prefix representing the digits in the atomic number, until the International Union of Pure and Applied Chemistry (IUPAC) approves a trivial name.

1. The atomic number of an atom is always equal to the number of its

   (1) protons, only
   (2) neutrons, only
   (3) protons plus neutrons
   (4) protons plus electrons   1 _____

2. The modern model of the atom shows that electrons are

   (1) orbiting the nucleus in fixed paths
   (2) found in regions called orbitals
   (3) combined with neutrons in the nucleus
   (4) located in a solid sphere covering the nucleus   2 _____

3. Which two particles each have a mass approximately equal to one atomic mass unit?

   (1) electron and neutron
   (2) electron and positron
   (3) proton and electron
   (4) proton and neutron   3 _____

4. Which diagram represents the nucleus of an atom of $^{27}_{13}\text{Al}$?

   14 n
   27 p

   (1)

   27 n
   13 p

   (3)

   14 n
   13 p

   (2)

   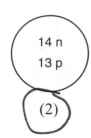
   40 n
   13 p

   (4)   4 _____

5. Which statement is true about the charges assigned to an electron and a proton?

   (1) Both an electron and a proton are positive.
   (2) An electron is positive and a proton is negative.
   (3) An electron is negative and a proton is positive.
   (4) Both an electron and a proton are negative.   5 _____

6. Compared to a proton, an electron has

   (1) a greater quantity of charge and the same sign
   (2) a greater quantity of charge and the opposite sign
   (3) the same quantity of charge and the same sign
   (4) the same quantity of charge and the opposite sign   6 _____

7. Atoms of the same element that have different numbers of neutrons are classified as

   (1) charged atoms
   (2) charged nuclei
   (3) isomers
   (4) isotopes   7 _____

8. The charge of a beryllium-9 nucleus is

   (1) +13   (3) +5
   (2) +9   (4) +4   8 _____

9. All the isotopes of a given atom have

(1) the same mass number and the same atomic number
(2) the same mass number but different atomic numbers
(3) different mass numbers but the same atomic number
(4) different mass numbers and different atomic numbers

9 _____

10. The elements in Period 5 on the Periodic Table are arranged from left to right in order of

(1) decreasing atomic mass
(2) decreasing atomic number
(3) increasing atomic mass
(4) increasing atomic number

10 _____

11. The atomic mass of an element is calculated using the

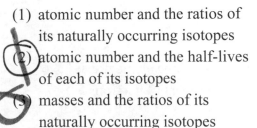

(1) atomic number and the ratios of its naturally occurring isotopes
(2) atomic number and the half-lives of each of its isotopes
(3) masses and the ratios of its naturally occurring isotopes
(4) masses and the half-lives of each of its isotopes

11 _____

12. In which shell are the valence electrons of the elements in Period 2 found?

(1) 1   (3) 3
(2) 2   (4) 4

12 _____

13. Which notation represents an atom of sodium with an atomic number of 11 and a mass number of 24?

(1) $^{24}_{11}Na$ (3) $^{13}_{11}Na$
(2) $^{11}_{24}Na$ (4) $^{35}_{11}Na$

13 _____

14. What is the total charge of the nucleus of a carbon atom?

(1) –6   (3) +6
(2) 0   (4) +12

14 _____

15. What is the total number of valence electrons in an atom of Mg-26 in the ground state?

(1) 1   (3) 3
(2) 2   (4) 4

15 _____

16. What is the total number of neutrons in an atom of $^{57}_{26}Fe$?

(1) 26   (3) 57
(2) 31   (4) 83

16 _____

17. What is the charge of the nucleus in an atom of oxygen-17?

(1) 0   (3) +8
(2) –2   (4) +17

17 _____

18. Compared to an atom of phosphorus-31, an atom of sulfur-32 contains

(1) one less neutron
(2) one less proton
(3) one more neutron
(4) one more proton

18 _____

19. The atomic mass unit is defined as exactly $\frac{1}{12}$ the mass of an atom of

(1) $^{12}_{6}C$　　　(3) $^{24}_{12}Mg$

(2) $^{14}_{6}C$　　　(4) $^{26}_{12}Mg$　　19 ____

20. Which element forms a compound with chlorine with the general formula MCl?

(1) Rb　　　(3) Re

(2) Ra　　　(4) Rn　　20 ____

21. In comparison to an atom of $^{19}_{9}F$ in the ground state, an atom of $^{12}_{6}C$ in the ground state has

(1) three fewer neutrons

(2) three fewer valence electrons

(3) three more neutrons

(4) three more valence electrons　　21 ____

Base your answers to question 22 on the information below.

In the modern model of the atom, each atom is composed of three major subatomic (or fundamental) particles.

22. *a)* Name the subatomic particles contained in the nucleus of the atom.

Protons and neutrons

b) State the charge associated with each type of subatomic particle contained in the nucleus of the atom. Protons-positive   neutrons-neutral (no charge)

c) What is the net charge of the nucleus? positive

23. State, in terms of subatomic particles, how an atom of C-13 is different from an atom of C-12.

It has one more proton than C-12
#of neutrons is diff

24. Explain, in terms of atomic particles, why S-32 is a stable nuclide.

# of neutrons + protons is equal

25. In the early 1900s, experiments were conducted to determine the structure of the atom. One of these experiments involved bombarding gold foil with alpha particles. Most alpha particles passed directly through the foil. Some, however, were deflected at various angles. Based on this alpha particle experiment, state two conclusions that were made concerning the structure of an atom.

Conclusion 1 _____

Conclusion 2 _____

Base your answers to question 26 on the data table below, which shows three isotopes of neon.

Isotope	Atomic Mass (atomic mass units)	Percent Natural Abundance
^{20}Ne	19.99	90.9%
^{21}Ne	20.99	0.3%
^{22}Ne	21.99	8.8%

of #neutrons

26. *a)* In terms of atomic particles, state one difference between these three isotopes of neon.

They have different mass #'s

b) Based on the atomic masses and the natural abundances shown in the data table, in the space below, show a correct numerical setup for calculating the average atomic mass of neon.

c) Based on natural abundances, the average atomic mass of neon is closest to which whole number? __20__

27. In the 19th century, Dmitri Mendeleev predicted the existence of a then unknown element X with a mass of 68. He also predicted that an oxide of X would have the formula X_2O_3. On the modern Periodic Table, what is the group number and period number of element X?

Group number_____ Period number_____

Base your answers to question 28 on the information below.

   The nucleus of one boron atom has five protons and four neutrons.

28. *a)* Determine the total number of electrons in the boron atom. ____5____

b) Determine the total charge of the boron nucleus. ____+5____

29. In the space below, draw a Lewis electron-dot diagram of a selenium atom in the ground state.

·S̈e̤:

Periodic Table

30. What determines the order of placement of the elements on the modern Periodic Table?

   (1) atomic number
   (2) atomic mass
   (3) the number of neutrons, only
   (4) the number of neutrons
       and protons   30 _____

31. A sample composed only of atoms having the same atomic number is classified as

   (1) a compound   (3) an element
   (2) a solution   (4) an isomer   31 _____

32. Which statement explains why sulfur is classified as a Group 16 element?

   (1) A sulfur atom has 6 valence electrons.
   (2) A sulfur atom has 16 neutrons.
   (3) Sulfur is a yellow solid at STP.
   (4) Sulfur reacts with most metals.   32 _____

33. What is the structure of a krypton-85 atom?

   (1) 49 electrons, 49 protons, and 85 neutrons
   (2) 49 electrons, 49 protons, and 49 neutrons
   (3) 36 electrons, 36 protons, and 85 neutrons
   (4) 36 electrons, 36 protons, and 49 neutrons
                                        33 _____

34. Which of these elements has an atom with the most stable outer electron configuration?

   (1) Ne   (3) Ca
   (2) Cl   (4) Na   34 _____

35. Which set of symbols represents atoms with valence electrons in the same electron shell?

   (1) Ba, Br, Bi   (3) O, S, Te
   (2) Sr, Sn, I   (4) Mn, Hg, Cu   35 _____

36. What are the characteristics of a neutron?

   (1) It has no charge and no mass.
   (2) It has no charge and a mass of 1 u.
   (3) It has a charge of +1 and no mass.
   (4) It has a charge of +1 and a mass
       of 1 u.   36 _____

37. An atom is electrically neutral because the

   (1) number of protons equals the
       number of electrons
   (2) number of protons equals the
       number of neutrons
   (3) ratio of the number of neutrons
       to the number of electrons is 1:1
   (4) ratio of the number of neutrons
       to the number of protons is 2:1   37 _____

38. In which group of the Periodic Table do most of the elements exhibit both positive and negative oxidation states?

   (1) 17   (3) 12
   (2) 2   (4) 7   38 _____

39. An atom of potassium-37 and an atom of potassium-42 differ in their total number of

   (1) electrons   (3) protons
   (2) neutrons   (4) positrons   39 _____

40. Which electron-dot symbol represents an atom of argon in the ground state?

(1) Ar: (3) ·Ar:

(2) ·Ar: (4) :Ar:

40 _____

41. Which statement concerning elements is true?

(1) Different elements must have different numbers of isotopes.
(2) Different elements must have different numbers of neutrons.
(3) All atoms of a given element must have the same mass number.
(4) All atoms of a given element must have the same atomic number.

41 _____

42. The atomic mass of an element is the weighted average of the masses of

(1) its two most abundant isotopes
(2) its two least abundant isotopes
(3) all of its naturally occurring isotopes
(4) all of its radioactive isotopes

42 _____

43. Subatomic particles can usually pass undeflected through an atom because the volume of an atom is composed of

(1) an uncharged nucleus
(2) largely empty space
(3) neutrons
(4) protons

43 _____

44. The mass of 12 protons is approximately equal to

(1) 1 atomic mass unit
(2) 12 atomic mass units
(3) the mass of 1 electron
(4) the mass of 12 electrons

44 _____

45. Which Lewis electron-dot diagram represents an atom in the ground state for a Group 13 element?

(1) :X: (3) X·

(2) X: (4) ·X:

40 _____

46. As the elements in Group 17 are considered in order of increasing atomic number, the chemical reactivity of each successive element

(1) decreases   (3) remains the same
(2) increases

46 _____

47. A student constructs a model for comparing the masses of subatomic particles. The student selects a small, metal sphere with a mass of 1 gram to represent an electron. A sphere with which mass would be most appropriate to represent a proton?

(1) 1 g   (3) $\frac{1}{2000}$ g

(2) $\frac{1}{2}$ g   (4) 2000 g

47 _____

48. A 100.00-gram sample of naturally occurring boron contains 19.78 grams of boron-10 (atomic mass = 10.01 atomic mass units) and 80.22 grams of boron-11 (atomic mass = 11.01 atomic mass units).

Which numerical setup can be used to determine the atomic mass of naturally occurring boron?

(1) $(0.1978)(10.01) + (0.8022)(11.01)$
(2) $(0.8022)(10.01) + (0.1978)(11.01)$
(3) $\frac{(0.1978)(10.01)}{(0.8022)(10.01)}$
(4) $\frac{(0.8022)(10.01)}{(0.1978)(10.01)}$

48 _____

49. Which statement describes the relative energy of the electrons in the shells of a calcium atom?

(1) An electron in the first shell has more energy than an electron in the second shell.
(2) An electron in the first shell has the same amount of energy as an electron in the second shell.
(3) An electron in the third shell has more energy than an electron in the second shell.
(4) An electron in the third shell has less energy than an electron in the second shell.

49 _____

50. According to the electron-cloud model of the atom, an orbital is a

(1) circular path traveled by an electron around the nucleus
(2) spiral path traveled by an electron toward the nucleus
(3) region of the most probable proton location
(4) region of the most probable electron location

50 _____

51. The most common isotope of chromium has a mass number of 52. Which notation represents a different isotope of chromium?

(1) $^{52}_{24}Cr$ (3) $^{24}_{52}Cr$
(2) $^{54}_{24}Cr$ (4) $^{24}_{54}Cr$

51 _____

52. Which statement identifies the element arsenic?
(1) Arsenic has an atomic number of 33.
(2) Arsenic has a melting point of 84 K.
(3) An atom of arsenic in the ground state has eight valence electrons.
(4) An atom of arsenic in the ground state has a radius of 146 pm.

52 _____

53. The table below gives information about the nucleus of each of four atoms.

Nuclei of Four Atoms

Atom	Number of Protons	Number of Neutrons
A	6	6
D	6	7
E	7	7
G	7	8

How many different elements are represented by the nuclei in the table?

(1) 1   (3) 3
(2) 2   (4) 4

53 _____

54. In the formula X_2O_5, the symbol X could represent an element in Group

(1) 1   (3) 15
(2) 2   (4) 18

54 _____

55. Each diagram below represents the nucleus of a different atom.

D   E   Q   R

Which diagrams represent nuclei of the same element?

(1) D and E, only   (3) Q and R, only
(2) D, E, and Q   (4) Q, R, and E

55 _____

56. If an element, X, can form an oxide that has the formula X_2O_3, then element X would most likely be located on the Periodic Table in the same group as

(1) Ba   (3) In
(2) Cd   (4) Na

56 _____

Base your answers to question 57 on the information below.

An atom has an atomic number of 9, a mass number of 19, and an electron configuration of 2–7.

57. *a*) What is the total number of neutrons in this atom? _____

b) What is the charge of this atom? _____

The accompanying chart gives information about two isotopes of element *X*.

58. Calculate the average atomic mass of element *X*.

Isotope	Mass	Relative Abundance
X-10	10.01	19.91%
X-11	11.01	80.09%

- Show a correct numerical setup in the space provided below.
- Record your answer.
- Express your answer to the correct number of significant figures.

59. In the accompanying space, draw a Lewis electron-dot diagram for a sulfur atom in the ground state.

Base your answers to question 60 on the information below, which describes the proposed discovery of element 118.

In 1999, a nuclear chemist and his team announced they had discovered a new element by crashing krypton atoms into lead. The new element, number 118, was assigned the name ununoctium and the symbol Uuo. One possible isotope of ununoctium could have been Uuo-291. However, the discovery of Uuo was not confirmed because other scientists could not reproduce the experimental results published by the nuclear chemist and his team. In 2006, another team of scientists claimed that they produced Uuo. This claim has yet to be confirmed. Adapted from Discover January 2002

60. *a*) Based on atomic number, in which group on the Periodic Table would element 118 be placed? _____

b) What would be the total number of neutrons present in a theoretical atom of Uuo-291? _____

c) What would be the total number of electrons present in a theoretical atom of Uuo-291? _____

Periodic Table

Periodic Table – Atomic Structure
Answers
Set 1

1. 1 The atomic number is defined as the number of protons in the nucleus of that element. All atoms of the same element have the same atomic number.

2. 2 Electrons, negatively charged particles, are found outside the nucleus. They are located in orbitals or energy levels that are at different locations around the nucleus. These orbitals are the regions where electrons are most likely to be found. This modern model is called the atomic orbital model.

3. 4 The nucleus consists of two subatomic particles, protons and neutrons. Both of these particles have approximately the same mass, which is equal to one atomic mass unit.

4. 2 Aluminum has an atomic mass number of 27 and an atomic number of 13. The atomic number is the number of protons in a nucleus. Subtracting the atomic number from the atomic mass gives the number of neutrons, in this case, 14.

5. 3 Electrons are negatively charged particles found in energy levels outside the nucleus of the atom. Protons are positively charged particles found in the nucleus.

6. 4 Electrons are negatively charged particles and protons are positively charged particles. Their charges are –1 and +1, respectively. Therefore, they have the same quantity of charge but opposite sign.

7. 4 By definition, isotopes are atoms having the same atomic number (number of protons) but having different mass numbers (number of protons and neutrons). Therefore, isotopes of an element will have different numbers of neutrons.

8. 4 The nucleus of atoms consists of protons and neutrons. A proton possesses a unit positive charge (+1), while a neutron has no charge (0). Thus, the nucleus is always positively in charged. As shown in the Periodic Table, the beryllium atom has an atomic number of 4, which is its number of protons. The nucleus of a beryllium atom would have a total charge of +4.

9. 3 Atoms of the same element have the same atomic number. Some atoms of the same element have different numbers of neutrons, thus differing in mass number. Atoms that have the same number of protons but different numbers of neutrons are called isotopes.

10. 4 Open to the Periodic Table and locate the elements in Period 5. The atomic number increases by one from one element to the next throughout the period. Notice the decrease in mass going from Te to I.

11. 3 The atomic mass is defined as the weighted average mass of the naturally occurring isotopes of that element. Isotopes of an element have the same atomic number but different numbers of neutrons resulting in different masses for each isotope.

12. 2 Open to the Periodic Table. All of the elements in Period 2 have two electron shells. The valence electrons are the electrons found in the last electron shell. For Period 2, this is the second electron shell. Remember, the number of the valence shell is usually the same as the period number of the element.

13. 1 In the notation for an element, the mass number is placed in the upper left of the symbol and the atomic number is placed in the lower left of the symbol.

14. 3 The nucleus of all atoms consists of protons and neutrons. A proton possesses a unit positive charge (+1), while a neutron has no charge (0). Thus the nucleus is always positively charged. As shown in the Periodic Table, the carbon atom has an atomic number of 6, which is its number of protons. The nucleus of a carbon atom would have a total charge of +6.

15. 2 An atom in the ground state would have all of its electrons occupying the lowest energy levels available. The Periodic Table shows Mg in the ground state with the electronic configuration of 2-8-2. By definition, the valence shell is the outermost electron shell in an atom. Mg-26, an isotope of Mg, contains two valence electrons.

16. 2 To obtain the number of neutrons, subtract the atomic number (number of protons) from the atomic mass number (number of protons and neutrons). Using this rule, $57 - 26 = 31$ neutrons.

17. 3 All isotopes of an element have the same atomic number. For oxygen, this is 8. This atomic number is the number of positively charged protons that are located in the nucleus. The neutron, which is also located in the nucleus, has no charge. The charge of the nucleus on O-17 is +8.

18. 4 Open to the Periodic Table and locate P (phosphous-31) and S (sulfur-32). Phosphorus has an atomic number of 15, thus its nucleus would contain15 protons. Sulfur has an atomic number of 16, its nucleus would contain 16 protons. They both have the same number of neutrons, 16.

19. 1 As shown on the Key to the Periodic Table of the elements, relative atomic masses are based on the ^{12}C atom. This atom is assigned a mass of exactly 12.00 atomic mass units (u). Therefore, 1 u is $\frac{1}{12}$ the mass of the standard ^{12}C atom.

20. 1 In all compounds, the total oxidation numbers must equal zero. The Periodic Table shows the oxidation number for chlorine is −1. In the formula MCl, element M must have an oxidation number of +1. All elements in Group 1, having an oxidation number of +1, can react with Cl to produce a compound with the general formula of MCl. Rb, located in Group 1, reacts with Cl producing the compound RbCl.

21. 2   An atom in the ground state would have all of its electrons occupying the lowest energy levels available. The Periodic Table shows the ground state configuration of fluorine and carbon as 2-7 and 2-4, respectively. Therefore, carbon would have 3 fewer valence electrons compared to fluorine. Fluorine has 10 neutrons (19 – 9 = 10), while carbon has 6 neutrons (12 – 6 = 6).

22. *a*) Answer: Protons and neutrons

Explanation: The nucleus of an atom is composed of protons and neutrons.

b) Answer: The proton has a positive charge (+), and the neutron has no charge (0).

Explanation: Through numerous experiments it has been proven that the proton has a positive charge while the neutron has a charge of zero, or neutral.

c) Answer: Positive (+)

Explanation: The nucleus consists of protons and neutrons. The protons are positive (+) in charge and the neutrons have no charge (0). This makes the nucleus of an atom positive in charge.

23.   Answer: A C-13 atom has seven neutrons and a C-12 atom has six neutrons.
or An atom of C-13 and an atom of C-12 have different numbers of neutrons.
or The number of neutrons is different.

Explanation: C-12 and C-13 are isotopes of carbon, having different numbers of neutrons. Open the Periodic Table and locate C. Subtracting the atomic number (6) from its atomic mass number (12) gives 6 for the number of neutrons for C-12. C-13, with an atomic mass number of 13 and an atomic number of 6, would have 7 neutrons.

24.   Answer: The neutron-to-proton ratio causes the nuclide to be stable. *or* The nuclide has an equal number of neutrons and protons. *or* Because of neutron-protons ratio are equal.

Explanation: As show by the Periodic Table, S-32 contains 16 protons and 16 neutrons. This makes a 1:1 ratio between the number of protons and neutrons. This ratio makes the nucleus of an atom stable.

25.   Answer: The nucleus is small. *or* The nucleus is positively charged.
or The atom is mostly empty space. *or* The nucleus is dense.

Explanation: The atom, consisting of a central nucleus composed of protons and neutrons with electrons in orbitals outside the nucleus, has been experimentally proven to consist of mostly empty space. Lord Ernest Rutherford, using subatomic particles (alpha particles) directed at a thin gold foil observed that almost all of the particles passed through the foil, showing the "emptiness" of the atom. The few deflected alpha particles indicated a dense positively charged nucleus in the center of the atom repelling the positively charged alpha particles.

26. *a*) Answer: Each isotope has a different number of neutrons.
or These isotopes have different number of neutrons.
or Ne-22 has two more neutrons than Ne-20 and one more neutron than Ne-21.
or Each isotope has a different mass number.

Explanation: Isotopes of the same element have the same atomic number but different mass numbers. In this chart, notice the different mass numbers of the three isotopes of Ne. Thus, isotopes of an element will have different numbers of neutrons. Remember, to get the number of neutrons of an isotope, subtract the atomic number from the atomic mass number.

b) Answer: $(0.909)(19.99) + (0.003)(20.99) + (0.088)(21.99)$

or $(90.9\%)(19.99) + (0.3\%)(20.99) + (8.8\%)(21.99)$

or $\dfrac{(90.9)(19.99) + (0.3)(20.99) + (8.8)(21.99)}{100}$

Explanation: The atomic mass of an element is the weighted average mass of the naturally occurring isotopes of that element. This is obtained by taking the sum of the products of the percentages and masses of each isotope. The above examples are the mathematical set-ups to obtain the correct average atomic mass of neon.

c) Answer: 20

Explanation: Based on the definition of atomic mass, the mass number of that element will be that of the most abundant element. From the chart, over 90% of all isotopes of neon are ^{20}Ne with an atomic mass of 19.99. This rounds off to 20.

27. Answer: Group 13 and Period 4

Explanation: The atomic mass of 68 would be located between Zn and Ge. This places element X in Period 4. The oxidation state of unknown element X is +3 as determined from the formula X_2O_3. In all formulas, the sum of the oxidation states must equal zero. The Periodic Table shows oxygen having an oxidation state of –2. With three oxygen atoms there would be a total oxidation state of –6. With two atoms of X, each X must have an oxidation state of +3, to make a total of +6. All elements in Group 13 have an oxidation state of +3.

28. a) Answer: 5 or five

Explanation: In an neutral atom the electrons equal the protons.

b) Answer: +5

Explanation: The nucleus contains protons and neutrons. Protons are positive in charge, while neutrons have no charge. This makes the charge of the nucleus of any atom positive, equal to the number of protons for that atom.

29. Answer: $\cdot \ddot{\underset{\cdot}{Se}} \colon$ Note – the location of electrons (dots) may vary.

Explanation: The Lewis electron-dot diagram consists of the symbol of the element surrounded by dots representing valence electrons. The Periodic Table shows Se has 6 valence electrons, thus the Lewis diagram must have 6 dots surrounding the symbol. Usually a maximum of two dots are positioned on each side of the symbol.

1. Which is a property of most nonmetallic solids?

   (1) high thermal conductivity
   (2) high electrical conductivity
   (3) brittleness
   (4) malleability

   1 _____

2. Which of the following Group 15 elements has the greatest metallic character?

   (1) nitrogen   (3) antimony
   (2) phosphorus   (4) bismuth

   2 _____

3. Which pair of symbols represents a metalloid and a noble gas?

   (1) Si and Bi   (3) Ge and Te
   (2) As and Ar   (4) Ne and Xe

   3 _____

4. Which group contains a metalloid?

   (1) 1   (3) 15
   (2) 11   (4) 18

   4 _____

5. Which element is malleable and can conduct electricity in the solid phase?

   (1) iodine   (3) sulfur
   (2) phosphorus   (4) tin

   5 _____

6. At STP, which element is brittle and not a conductor of electricity?

   (1) S   (3) Na
   (2) K   (4) Ar

   6 _____

7. Element X is a solid that is brittle, lacks luster, and has six valence electrons. In which group on the Periodic Table would element X be found?

   (1) 1   (3) 15
   (2) 2   (4) 16

   7 _____

8. An element that is malleable and a good conductor of heat and electricity could have an atomic number of

   (1) 16   (3) 29
   (2) 18   (4) 35

   8 _____

9. Which list of elements contains a metal, a metalloid, and a nonmetal?

   (1) Zn, Ga, Ge
   (2) Si, Ge, Sn
   (3) Cd, Sb, I
   (4) F, Cl, Br

   9 _____

A metal, M, was obtained from a compound in a rock sample. Experiments have determined that the element is a member of Group 2 on the Periodic Table of the Elements.

10. a) What is the phase of element M at STP? ~~A nonmetal~~ solid

    b) Give 3 characteristic properties of metals. 1) malleable

    2) good conductor of e

    3) good thermal con

11. The elements located in the lower left corner of the Periodic Table are classified as

    (1) metals   (3) metalloids
    (2) nonmetals   (4) noble gases   11 _____

12. As the atomic number of elements within Group 2 increases, the metallic character of each successive element

    (1) decreases
    (2) increases
    (3) remains the same   12 _____

13. Which element has both metallic and nonmetallic properties?

    (1) Rb   (3) Si
    (2) Rn   (4) Sr   13 _____

14. In which compound is the ratio of metal ions to nonmetal ions 1 to 2?

    (1) calcium bromide
    (2) calcium oxide
    (3) calcium phosphide
    (4) calcium sulfide   14 _____

15. Arsenic and silicon are similar in that they both

    (1) have the same ionization energy
    (2) have the same covalent radius
    (3) are transition metals
    (4) are metalloids   15 _____

16. Which element is an alkali metal?

    (1) hydrogen   (3) sodium
    (2) calcium   (4) zinc   16 _____

17. The element in Period 4 and Group 1 of the Periodic Table would be classified as a

    (1) metal   (3) nonmetal
    (2) metalloid   (4) noble gas   17 _____

18. Which Period 4 element has the most metallic properties?

    (1) As   (3) Ge
    (2) Br   (4) Sc   18 _____

19. Which element is classified as a nonmetal?

    (1) Be   (3) Si
    (2) Al   (4) Cl   19 _____

20. At STP, an element that is a solid and a good conductor of heat and electricity could have an electron configuration of

    (1) 2-7   (3) 2-8-5
    (2) 2-8-8   (4) 2-8-18-2   20 _____

21. Which element is malleable and a good conductor of electricity at STP?

    (1) argon   (3) iodine
    (2) carbon   (4) silver   21 _____

22. Describe one chemical property of Group 1 metals that results from the atoms of each metal having only one valence electron.

_____

23. Explain, in terms of atomic structure, why liquid mercury is a good electrical conductor.

_____

1. 3 Nonmetallic solids usually are brittle, tend to be nonconductors, show little or no luster, and are not malleable.

2. 4 Looking at the Periodic Table, locate the dark ziz-zag line. This line separates elements with metallic properties (to the left of the line) from the elements that have nonmetallic properties (to the right of the line). Bismuth is the only choice that is on the metallic side of the Periodic Table. Also, in general, metallic characteristics increase with an increase in the atomic number within a group on the Periodic Table.

3. 2 Metalloids are elements that exhibit properties of metals and nonmetals, also referred to as semimetals. Metalloids are found adjacent to the dark zig-zag line (especially to the right) on the Periodic Table. As (arsenic) is considered to be a metalloid. Ar (argon) is a Group 18 element, known as the noble or inert gases.

4. 3 Metalloids are located adjacent to the zig-zag line. Of the given choices only Group 15 contains an element that is adjacent to this line.

5. 4 Metals tend to exhibit the properties of malleability and conductivity. Tin is the only metal from the given choices. All others choices are nonmetals, which would not exhibit these properties.

6. 1 Nonmetallic solids are brittle and do not conduct electricity. Nonmetals are located on the right side of the zig-zag line on the Periodic Table. By the process of elimination Na and K are metals. Ar is an inert gas. This leaves sulfur (S) as the element that is brittle and a nonconductor of electricity.

7. 4 The properties of element X indicate that it is a nonmetal, therefore it must be located to the right of the zig-zag line. Possessing six valence electrons places it in Group 16.

8. 3 The properties given describes a metal. Atomic number 29 is the element copper, which is a metal, being is on the left side of the darken zig-zag line on the Periodic Table.

9. 3 Elements to the left of the zig-zag line found on the Periodic Table are metals. Elements to the right of this line are nonmetals. Metalloids are found adjacent to this zig-zag line. Answer 3 fulfills these requirements.

10. *a)* Answer : Solid

Explanation: Table A gives the values of STP. Using these values and Table S, all elements in Group 2 elements have a much higher melting point than 273 K, which make them solids at STP.

b) Answer: malleability *or* high electrical conductivity *or* high thermal conductivity *or* exhibiting luster or ductilility *or* ductility.

Explanation: The above properties are associated with metals.

1. Which element is most chemically similar to chlorine?

   (1) Ar   (3) Fr

   (2) F   (4) S   1____

2. Which list consists of elements that have the most similar chemical properties?

   (1) Mg, Al, and Si

   (2) Mg, Ca, and Ba

   (3) K, Al, and Ni

   (4) K, Ca, and Ga   2____

3. Based on the Periodic Table, explain why Na and K have similar chemical properties.

   They are in the same group

4. Explain, in terms of electron configuration, why oxygen atoms and sulfur atoms form compounds with similar molecular structures.

   They both have 6 valence e-

Base your answers to question 5 on the information below.

Given: Samples of Na, Ar, As, Rb

5. a) Which two of the given elements have the most similar chemical properties?

   1)___Na___   2)___Rb___

   b) Explain your answer in terms of the Periodic Table of the Elements.

   They are in the same group

6. Arsenic and bismuth are similar in that they both

    (1) have the same ionization energy
    (2) have the same number of valence electrons
    (3) are transition metals
    (4) are metalloids   6 _____

7. Which element has chemical properties that are most similar to those of calcium?

    (1) Co   (3) N
    (2) K   (4) Sr   7 _____

8. Which element has chemical properties that are most similar to the chemical properties of sodium?

    (1) Mg   (3) Se
    (2) K   (4) Cl   8 _____

9. As the atoms of the Group 17 elements in the ground state are considered from top to bottom, each successive element has

    (1) the same number of valence electrons and similar chemical properties
    (2) the same number of valence electrons and identical chemical properties
    (3) an increasing number of valence electrons and similar chemical properties
    (4) an increasing number of valence electrons and identical chemical properties   9 _____

10. Explain, in terms of electron configuration, why selenium and sulfur have similar chemical properties.

    _____

    _____

11. In the space below, give the symbol of an element that has similar chemical properties as the one shown below and draw in the Lewis electron-dot diagram for this element in the ground state.

$\overset{\displaystyle ..}{Al} \cdot$

1. 2   Locate chlorine (Cl) in Group 17. Elements within a group have similar chemical properties. This similarity is due to the electron structure of the elements within a group, each containing the same number of valence electrons. Of the choices given, only fluorine (F) is in the same group as chlorine (Cl), both having 7 valence electrons that gives them similar chemical properties.

2. 2   Mg, Ca, and Ba are all located in Group 2. These metallic elements all have similar electron structures, each containing 2 valence electrons. Therefore, they will exhibit similar chemical properties.

3.   Answer: They have the same number of valence electrons

   or   form 1^+ ions

   or   are located in same group

   or   both alkali metals

   Explanation: Locate sodium (Na) and potassium (K) in Group 1, the alkali metals. The elements within a group have similar chemical properties, due to the electron structure of the elements within the group. Elements in this group have one valence electron and will readily give up this electron in similar chemical reactions.

4.   Answer: Oxygen and sulfur atoms have the same number of valence electrons (6).

   or   Atoms of both elements need two more valence electrons to complete their outer shells.

   Explanation: Sulfur and oxygen are both in Group 16. Elements in the same group have the same number of valence electron numbers, in this case 6. This causes the elements in the same group to form compounds with similar molecular structures and undergo similar reactions.

5. *a)* Answer: Na and Rb

   b) Explanation: Sodium (Na) and rubidium (Rb) are both in Group 1. The elements within a group have similar chemical properties due to the electron structure of the elements within the group. Both Na and Rb have one valence electron and will readily give this electron up in similar chemical reactions.

Adde-

1. What is the total number of electrons in a Cu^+ ion?

   (1) 28   (3) 30   ✓
   (2) 29   (4) 36   1 _____

2. What is the total number of electrons in a Mg^{2+} ion?

   (1) 10   (3) 14
   (2) 12   (4) 24   2 _____

3. What is the total number of electrons in a S^{2-} ion?

   (1) 10   (3) 16 +2
   (2) 14   (4) 18   3 _____

4. Which ion has the same electron configuration as an atom of He?

   (1) H^- (3) Na^+
   (2) O^{2-} (4) Ca^{2+} ✓   4 _____

5. Which symbol represents a particle that has the same total number of electrons as S^{2-}?

   (1) O^{2-} (3) Se^{2-}
   (2) Si   (4) Ar   16   5 _____

6. Which change in oxidation number indicates oxidation? LEO

   (1) 1^- to 2^+ (3) 2^+ to 3^-
   (2) 1^- to 2^- (4) 3^+ to 2^+ 6 _____

7. How many electrons are contained in an Au^{3+} ion?

   (1) 76   (3) 82
   (2) 79   (4) 197   7 _____

8. Which electron configuration could represent a strontium atom in an excited state?

   (1) 2–8–18–7–1 (3) 2–8–18–8–1   ✓
   (2) 2–8–18–7–3 (4) 2–8–18–8–2   8 _____

9. What is the net charge on an ion that has 9 protons, 11 neutrons, and 10 electrons?

   (1) 1+   (3) 1–
   (2) 2+   (4) 2–   ✓   9 _____

10. What is the total number of electrons in a Cr^{3+} ion? -3

   (1) 18   (3) 24
   (2) 21   (4) 27   10 _____

11. Which electron configuration represents the electrons of an atom in an excited state?

   (1) 2–4   (3) 2–7–2
   (2) 2–6   (4) 2–8–2   ✓   11 _____

12. Which changes occur as a cadmium atom, Cd, becomes a cadmium ion, Cd^{2+}? -2

   (1) The Cd atom gains two electrons and its radius decreases.
   (2) The Cd atom gains two electrons and its radius increases.
   (3) The Cd atom loses two electrons and its radius decreases.
   (4) The Cd atom loses two electrons and its radius increases.   12 _____

13. Which Lewis electron-dot diagram is correct for a S^{2-} ion?

(1) $[\cdot \overset{\cdot\cdot}{S} \cdot]^{2-}$

(2) $[\overset{\cdot\cdot}{S}]^{2-}$

(3) ⊚ $[\overset{\cdot\cdot}{:S} \cdot]^{2-}$ ✓

(4) $[\overset{\cdot\cdot}{:S:}]^{2-}$

13 ____

14. The light emitted from a flame is produced when electrons in an excited state

(1) absorb energy as they move to lower energy states

(2) ~~absorb energy as they move to higher energy states~~ ⊘

(3) release energy as they move to lower energy states

(4) ⊚ release energy as they move to higher energy states

14 ____

Base your answers to question 15 on the following redox reaction, which occurs spontaneously in an electrochemical cell.

$$Zn + Cr^{3+} \rightarrow Zn^{2+} + Cr$$

15. a) State what happens to the number of protons in a Zn atom when it changes to Zn^{2+} as the redox reaction occurs.

stays same

b) State what happens to the number of electrons in a Zn atom when it changes to Zn^{2+} as the redox reaction occurs.

loses 2 e-

Base your answers to question 16 on the accompanying electron configuration table.

Element	Electron Configuration
X	2–8–8–2
Y	2–8–7–3
Z	2–8–8

✓ 16. a) What is the total number of valence electrons in an atom of electron configuration X?

2

b) Which electron configuration represents an excited state of an atom? *2-8-7-3*

c) Give the name of Element X. *Ca*

17. Explain how a bright-line spectrum is produced, in terms of excited state, energy transitions, and ground state.

As e- move from Ground state to excited state, energy is released and light is produced

18. As an atom becomes an ion, its mass number

    (1) decreases
    (2) increases
    (3) remains the same   18 _____

19. Which symbol represents a particle with a total of 10 electrons?

    (1) N   (3) Al
    (2) N^{3+} (4) Al^{3+} 19 _____

20. Which half-reaction correctly represents reduction?

    (1) $Ag \rightarrow Ag^+ + e^-$
    (2) $F_2 \rightarrow 2\,F^- + 2e^-$
    (3) $Au^{3+} + 3e^- \rightarrow Au$
    (4) $Fe^{2+} + e^- \rightarrow Fe^{3+}$ 20 _____

21. As a Ca atom undergoes oxidation to Ca^{2+}, the number of neutrons in its nucleus

    (1) decreases
    (2) increases
    (3) remains the same   21 _____

22. When a neutral atom undergoes oxidation, the atom's oxidation state

    (1) decreases as it gains electrons
    (2) decreases as it loses electrons
    (3) increases as it gains electrons
    (4) increases as it loses electrons   22 _____

23. Which group on the Periodic Table of the Elements contains elements that react with oxygen to form compounds with the general formula X_2O?

    (1) Group 1   (3) Group 14
    (2) Group 2   (4) Group 18   23 _____

24. Which species does not have a noble gas electron configuration?

    (1) Na^+ (3) Ar
    (2) Mg^{2+} (4) S   24 _____

25. Which Lewis electron-dot diagram correctly represents a hydroxide ion?

    (1) $\left[:\overset{..}{\underset{..}{O}}:H\right]^-$ (3) $\left[:\overset{..}{\underset{..}{O}}::H\right]^-$

    (2) $\left[:O:H:\right]^-$ (4) $\left[:O:\overset{..}{\underset{..}{H}}:\right]^-$ 25 _____

26. During which process does an atom gain one or more electrons?

    (1) transmutation
    (2) reduction
    (3) oxidation
    (4) neutralization   26 _____

27. Which particle has the same electron configuration as a potassium ion?

    (1) fluoride ion   (3) neon atom
    (2) sodium ion   (4) argon atom   27 _____

28. Given the cell reaction:

    $Ca(s) + Mg^{2+}(aq) \rightarrow Ca^{2+}(aq) + Mg(s)$

    Which substance is oxidized?

    (1) Ca(s)   (3) $Ca^{2+}(aq)$
    (2) $Mg^{2+}(aq)$ (4) Mg(s)   28 _____

29. Which electron configuration represents an atom of aluminum in an excited state?

    (1) 2-7-4   (3) 2-8-3
    (2) 2-7-7   (4) 2-8-6   29 _____

30. Which electron transition represents a gain of energy?

(1) from 2nd to 3rd shell
(2) from 2nd to 1st shell
(3) from 3rd to 2nd shell
(4) from 3rd to 1st shell   30 ____

31. What occurs when an atom of chlorine forms a chloride ion?

(1) The chlorine atom gains an electron, and its radius becomes smaller.
(2) The chlorine atom gains an electron, and its radius becomes larger.
(3) The chlorine atom loses an electron, and its radius becomes smaller.
(4) The chlorine atom loses an electron, and its radius becomes larger.
                                   31 ____

32. Compared to a sodium atom in the ground state, a sodium atom in the excited state must have

(1) a greater number of electrons
(2) a smaller number of electrons
(3) an electron with greater energy
(4) an electron with less energy   32 ____

33. During a flame test, ions of a specific metal are heated in the flame of a gas burner. A characteristic color of light is emitted by these ions in the flame when the electrons

(1) gain energy as they return to lower energy levels
(2) gain energy as they move to higher energy levels
(3) emit energy as they return to lower energy levels
(4) emit energy as they move to higher energy levels   33 ____

34. Write one electron configuration for an atom of silicon in an excited state. _____

35. Write an electron configuration for an atom of aluminum-27 in an excited state. _____

Base your answers to question 36 on the information and the bright-line spectra below.

Many advertising signs depend on the production of light emissions from gas-filled glass tubes that are subjected to a high-voltage source. When light emissions are passed through a spectroscope, bright-line spectra are produced.

Gas A

Gas B

Gas C

Gas D

Unknown mixture

36. a) Identify the two gases in the unknown mixture.

1) _____ 2) _____

b) Explain the production of an emission spectrum in terms of the energy states of an electron.

_____

_____

Base your answer to question 37 on the information below.

Potassium ions are essential to human health. The movement of dissolved potassium ions, $K^+(aq)$, in and out of a nerve cell allows that cell to transmit an electrical impulse.

37. What is the total number of electrons in a potassium ion? _____

Base your answer to question 38 on your knowledge of chemical bonding and on the Lewis electron-dot diagrams of H_2S, CO_2, and F_2 below.

$$H:\overset{\cdot\cdot}{\underset{H}{S}}: \qquad :\overset{\cdot\cdot}{O}::C::\overset{\cdot\cdot}{O}: \qquad :\overset{\cdot\cdot}{\underset{\cdot\cdot}{F}}:\overset{\cdot\cdot}{\underset{\cdot\cdot}{F}}:$$

38. Which atom, when bonded as shown, has the same electron configuration as an atom of argon?

_____

Base your answer to question 39 on the diagram and information given below.

39. a) Give the symbol and the charge of an aluminum ion.

_____

b) Write the formula for the compound formed between aluminum and oxygen.

_____

c) Explain why Lewis electron-dot diagrams are generally more suitable than electron-shell diagrams for illustrating chemical bonding.

Atomic Diagrams of Magnesium and Aluminum

Element	Lewis Electron-Dot Diagram	Electron-Shell Diagram
magnesium	Mg:	(12 p, 11 n)
aluminum	Al:	(13 p, 14 n)

Key
● = electron

_____

_____

40. In the accompanying space, draw a Lewis electron-dot diagram for CF_4.

Periodic Table – Ions and Excited Atoms
Answers
Set 1

1. 1   The Periodic Table shows that the electron configuration of copper is 2-8-18-1, having a total of 29 electrons. To form the Cu^+ ion, the Cu atom loses its valence electron, leaving the atom with 28 electrons.

2. 1   The Periodic Table shows that the electron configuration of Mg^{2+} is 2-8-2 having a total of 12 electrons. In chemical reactions it will give up its two valence electrons. Thus, the resulting ion will be Mg^{2+} with an electron configuration of 2-8. This ion will now have 10 electrons.

3. 4   The Periodic Table shows that S has an electron configuration of 2-8-6. To form the S^{2-} ion, the S atom gains two electrons, giving it an electron configuration of 2-8-8. This ion will now have a total of 18 electrons.

4. 1   As shown in the Periodic Table, helium's electron configuration is 2. A neutral hydrogen atom will have the electron configuration of 1. If hydrogen gains one electron, the resulting ion will be H^- with an electron configuration of 2.

5. 4   The Periodic Table shows that the sulfur atom has an electron configuration of 2-8-6. If a sulfur atom gains two electrons, the resulting ion will be S^{2-} with an electron configuration of 2-8-8. This would give a total of 18 electrons for the S^{2-} ion. The neutral atom of Ar also has 18 electrons, as shown in the Periodic Table.

6. 1   In oxidation, the element loses an electron(s), resulting in an increase in the oxidation number. Only choice 1 shows an increase in oxidation number.

7. 1   Open to the Periodic Table and it shows that the atomic number for Au is 79. A neutral atom of gold would have 79 protons and 79 electrons. Au^{3+} has undergone oxidation in which there is a loss of three electrons. Therefore, the total number of electrons in the Au^{3+} ion is 76.

8. 2   In the ground state, strontium (Sr) has an electron configuration of 2-8-18-8-2 (see Periodic Table), for a total of 38 electrons. When an atom absorbs enough energy one or more electrons can move to a higher energy level. When this occurs the atom is considered to be in an excited state. Answer 2 has 38 electrons, but represents an excited state, as shown by the vacancy of an electron in the second to last shell caused by the electron moving up to the next energy level. Choice 1 and 3 do not contain 38 electrons.

9. 3   This ion has one more electron than the number of protons. This would give this ion a –1 charge. Neutrons have no charge.

10. 2   The Periodic Table shows that Cr has an electron configuration of 2-8-13-1 for a total of 24 electrons. If Cr^0 loses 3 electrons, the resulting ion would be Cr^{3+}, containing 21 electrons.

11. 3   An atom in the ground state would have all of its electrons occupying the lowest energy levels available. When an atom absorbs enough energy, one or more electrons can move to a higher energy level. When this occurs the atom is considered to be in an excited state. The electron configurations on the Periodic Table are ground state configurations. The Periodic Table shows sodium with an electron configuration of 2-8-1, for a total of 11 electrons. Choice 3 has 11 electrons but the second shell contains 7 electrons, not the expected 8. This missing electron, causing the vacancy, moved up to the next energy level, thus 2-7-2 represents the electrons of a sodium atom in the excited state.

12. 3   The Cd atom loses two electrons in forming the Cd^{2+} ion. By losing two electrons, the entire valence shell is lost (see Group 12), thus causing a decrease in radius of the atom.

13. 4   A sulfur atom has the electron configuration 2-8-6. To form the S^{2-} ion, the atom must gain two electrons, giving it the configuration 2-8-8. The Lewis electron-dot diagram consists of the symbol of the element surrounded by dots representing valence electrons. Choice 4 shows 8 dots representing the 8 valence electrons.

14. 3   At high temperatures, an atom may become excited. When this occurs, one or more electrons will move to a higher energy level. When this excited electron(s) returns to its lower energy level, it will emit energy in the form of light. When viewed through a spectroscope, spectral lines of that element will be observed.

15. *a)*   Answer: The number of protons remains the same.
     or   The number of protons is unchanged.

     Explanation: The protons are in the nucleus of the atom and are not affected by the chemical reactions involved. Chemical reactions involve only electrons.

   b)   Answer: It loses two electrons.
     or Zn^0 undergoes oxidation by losing two electrons.

     Explanation: When an element loses electrons its oxidation number will increase. In the oxidation of Zn^0 to Zn^{2+}, the atom loses two electrons.

16. *a*) Answer: 2

Explanation: Element *X* has the electronic configuration of 2-8-8-2. The valence shell is the last or outermost shell in this configuration.

b) Answer: Element *Y* or 2-8-7-3

Explanation: When an atom absorbs enough energy, one or more electrons can move to higher energy levels. When this occurs, the atom is considered to be in an excited state. The Periodic Table shows calcium (in the ground state) with an electron configuration of 2-8-8-2 for a total of 20 electrons. Element *Y* has 20 electrons but the third shell contains 7 electrons, not the expected 8. This missing electron, causing the vacancy, moved up to the next energy level. Therefore, element *Y* would represent a calcium atom in the excited state.

c) Answer: Calcium

Explanation: In all neutral atoms, protons equal electrons. Element *X* would have 20 protons or an atomic number of 20. From the Periodic Table this is calcium.

17.   Answer: Moving from the excited state to the ground state releases energy.
        or An electron absorbs energy and moves to a higher energy level. As the electron returns to a lower energy level, it releases energy in the form of a bright-line spectrum.

Explanation: When an atom gains enough energy, an electron is able to move to a higher energy level. When the electron returns to the ground state, energy is released as light. A spectroscope separates this light into bright spectral lines.

Atomic Number	Symbol	Name	First Ionization Energy (kJ/mol)	Electro- negativity	Melting Point (K)	Boiling* Point (K)	Density** (g/cm³)	Atomic Radius (pm)
1	H	hydrogen	1312	2.2	14	20.	0.000082	32
2	He	helium	2372	—	—	4	0.000164	37
3	Li	lithium	520.	1.0	454	1615	0.534	130.
4	Be	beryllium	900.	1.6	1560.	2744	1.85	99
5	B	boron	801	2.0	2348	4273	2.34	84
6	C	carbon	1086	2.6	—	—	—	75
7	N	nitrogen	1402	3.0	63	77	0.001145	71
8	O	oxygen	1314	3.4	54	90.	0.001308	64
9	F	fluorine	1681	4.0	53	85	0.001553	60.
10	Ne	neon	2081	—	24	27	0.000825	62
11	Na	sodium	496	0.9	371	1156	0.97	160.
12	Mg	magnesium	738	1.3	923	1363	1.74	140.
13	Al	aluminum	578	1.6	933	2792	2.70	124
14	Si	silicon	787	1.9	1687	3538	2.3296	114
15	P	phosphorus (white)	1012	2.2	317	554	1.823	109
16	S	sulfur (monoclinic)	1000.	2.6	388	718	2.00	104
17	Cl	chlorine	1251	3.2	172	239	0.002898	100.
18	Ar	argon	1521	—	84	87	0.001633	101
19	K	potassium	419	0.8	337	1032	0.89	200.
20	Ca	calcium	590.	1.0	1115	1757	1.54	174
21	Sc	scandium	633	1.4	1814	3109	2.99	159
22	Ti	titanium	659	1.5	1941	3560.	4.506	148
23	V	vanadium	651	1.6	2183	3680.	6.0	144
24	Cr	chromium	653	1.7	2180.	2944	7.15	130.
25	Mn	manganese	717	1.6	1519	2334	7.3	129
26	Fe	iron	762	1.8	1811	3134	7.87	124
27	Co	cobalt	760.	1.9	1768	3200.	8.86	118
28	Ni	nickel	737	1.9	1728	3186	8.90	117
29	Cu	copper	745	1.9	1358	2835	8.96	122
30	Zn	zinc	906	1.7	693	1180.	7.134	120.
31	Ga	gallium	579	1.8	303	2477	5.91	123
32	Ge	germanium	762	2.0	1211	3106	5.3234	120.
33	As	arsenic (gray)	944	2.2	1090.	—	5.75	120.
34	Se	selenium (gray)	941	2.6	494	958	4.809	118
35	Br	bromine	1140.	3.0	266	332	3.1028	117
36	Kr	krypton	1351	—	116	120.	0.003425	116
37	Rb	rubidium	403	0.8	312	961	1.53	215
38	Sr	strontium	549	1.0	1050.	1655	2.64	190.
39	Y	yttrium	600.	1.2	1795	3618	4.47	176
40	Zr	zirconium	640.	1.3	2128	4682	6.52	164

Atomic Number	Symbol	Name	First Ionization Energy (kJ/mol)	Electro-negativity	Melting Point (K)	Boiling* Point (K)	Density** (g/cm³)	Atomic Radius (pm)
41	Nb	niobium	652	1.6	2750.	5017	8.57	156
42	Mo	molybdenum	684	2.2	2896	4912	10.2	146
43	Tc	technetium	702	2.1	2430.	4538	11	138
44	Ru	ruthenium	710.	2.2	2606	4423	12.1	136
45	Rh	rhodium	720.	2.3	2237	3968	12.4	134
46	Pd	palladium	804	2.2	1828	3236	12.0	130.
47	Ag	silver	731	1.9	1235	2435	10.5	136
48	Cd	cadmium	868	1.7	594	1040.	8.69	140.
49	In	indium	558	1.8	430.	2345	7.31	142
50	Sn	tin (white)	709	2.0	505	2875	7.287	140.
51	Sb	antimony (gray)	831	2.1	904	1860.	6.68	140.
52	Te	tellurium	869	2.1	723	1261	6.232	137
53	I	iodine	1008	2.7	387	457	4.933	136
54	Xe	xenon	1170.	2.6	161	165	0.005366	136
55	Cs	cesium	376	0.8	302	944	1.873	238
56	Ba	barium	503	0.9	1000.	2170.	3.62	206
57	La	lanthanum	538	1.1	1193	3737	6.15	194
Elements 58–71 have been omitted.								
72	Hf	hafnium	659	1.3	2506	4876	13.3	164
73	Ta	tantalum	728	1.5	3290.	5731	16.4	158
74	W	tungsten	759	1.7	3695	5828	19.3	150.
75	Re	rhenium	756	1.9	3458	5869	20.8	141
76	Os	osmium	814	2.2	3306	5285	22.587	136
77	Ir	iridium	865	2.2	2719	4701	22.562	132
78	Pt	platinum	864	2.2	2041	4098	21.5	130.
79	Au	gold	890.	2.4	1337	3129	19.3	130.
80	Hg	mercury	1007	1.9	234	630.	13.5336	132
81	Tl	thallium	589	1.8	577	1746	11.8	144
82	Pb	lead	716	1.8	600.	2022	11.3	145
83	Bi	bismuth	703	1.9	544	1837	9.79	150.
84	Po	polonium	812	2.0	527	1235	9.20	142
85	At	astatine	—	2.2	575	—	—	148
86	Rn	radon	1037	—	202	211	0.009074	146
87	Fr	francium	393	0.7	300.	—	—	242
88	Ra	radium	509	0.9	969	—	5	211
89	Ac	actinium	499	1.1	1323	3471	10.	201
Elements 90 and above have been omitted.								

*boiling point at standard pressure
**density of solids and liquids at room temperature and density of gases at 298 K and 101.3 kPa
— no data available
Source: *CRC Handbook for Chemistry and Physics*, 91st ed., 2010–2011, CRC Press

Properties of Selected Elements

Overview:

This table shows some important chemical and physical properties of selected elements. These properties may be used to predict chemical reactivity, bond type, the phase of the element at a given temperature, the identity of an element and the ionic radius of an element.

The Table:

The *Atomic Number, Symbol* and *Name* of the selected elements are given in the first three columns. The elements are listed in order of increasing atomic number.

The *First Ionization Energy* is defined as the energy required to remove the most loosely bound (outermost) electron from an atom. This electron is an electron in the valence level of the atom. It is measured in kilojoules/mole (kJ/mol) of atoms.

The *Electronegativity* of an atom is the measure of the relative ability of that atom to attract bonding electrons to itself. It is measured relative to that of fluorine (F), assigned a value of 4.0, the highest electronegativity of all the elements. Since it is a relative measurement, there is no unit for electronegativity. Notice that there is no value listed for He, Ne, Ar, Kr and Rn. These elements are inert or noble gases and do not form chemical bonds with other elements. There is however a value listed for Xe, another inert gas, since Xe does form compounds with the most active elements, F and O.

The *Melting Point* is the temperature at which the solid changes to a liquid. Notice that the melting point is expressed on the Kelvin (K) temperature scale. The melting point and freezing point for an element are the same.

The *Boiling Point* is the temperature at which the liquid changes to the gas phase. It is expressed on the Kelvin scale also. Note at the bottom of the table that the boiling point is measured at standard pressure and is therefore the normal boiling point of the element. The boiling point and the condensation point for an element are the same.

The *Density* of a substance is defined as the mass per unit volume. It is expressed in units of gram/cubic centimeter (g/cm^3). Note at the bottom of the table that the density of solids and liquids are measured at room temperature and the densities of gases at 298 K and 101.3 kPa.

The *Atomic Radius* is an estimate of the size of an atom or the distance from the center of the nucleus to the edge of the atom. It is an estimate due to the fact the outer edge of an atom is not distinct. Atomic radii are measured in picometers (pm–see Table C).

Additional Information:

- The difference in electronegativity between two elements is used to determine the type of bond formed between those elements.
  - If the difference in electronegativity (ΔEN) is less than 1.7, the bond is a covalent bond.
  - If the ΔEN is between 0 and 1.7, the bond is a polar covalent bond.
  - If the ΔEN is 0, the bond is a nonpolar covalent bond.
  - If the ΔEN is greater than 1.7, the bond is ionic.

- The greater the ΔEN, the more polar the bond.

- Metals have relatively low ionization energies.
  Nonmetals have relatively high ionization energies.

- With increasing atomic number in a given period, the ionization energy generally increases.
  With increasing atomic number in a given group, the ionization energy generally decreases.

- As successive electrons are removed from an atom, the ionization energy increases.
  Thus the second ionization energy is greater than the first ionization energy,
  the third greater than the second and so forth.

- Metals have relatively low electronegativities.
  Nonmetals have relatively high electronegativities.

- With increasing atomic number in a given period, the electronegativity generally increases.
  With increasing atomic number in a given group, the electronegativity generally decreases.

- With increasing atomic number in a given period, the atomic radius generally decreases.
  With increasing atomic number in a given group, the atomic radius generally increases.

- When an atom loses an electron or electrons, its radius decreases. When an atom gains an electron or electrons, its radius increases.

- If an element is at a temperature below its melting point, it is a solid. At a temperature between its melting point and boiling point, it is a liquid. At a temperature above its boiling point, it is a gas or vapor.

- Density, melting point and boiling point are characteristic properties of an element and may be used to identify an unknown element.

— Set 1 —

1. The freezing point of bromine on the Kelvin temperature scale is

   (1) 332   (3) 490
   (2) 266   (4) 958   1_____

2. Which halogen is a solid at STP?

   (1) Br_2 (3) Cl_2
   (2) F_2 (4) I_2 2_____

3. At standard pressure, which element has a melting point higher than standard temperature?

   (1) F_2 (3) Fe
   (2) Br_2 (4) Hg   3_____

4. What is the total number of elements in Group 17 that are gases at standard temperature and standard pressure?

   (1) 1   (3) 3
   (2) 2   (4) 4   4_____

5. At STP, an element that is a brittle solid and a poor conductor of heat and electricity could have an atomic number of

   (1) 12   (3) 16
   (2) 13   (4) 17   5_____

6. At 300 K, in which phase of matter do most of the known elements exist?

   (1) solid   (3) gas
   (2) liquid   (4) supercooled liquid
       6_____

7. Which element has the greatest density?
   (1) copper   (3) mercury
   (2) lead   (4) gold   7_____

— Set 2 —

8. Which of these elements has the lowest melting point?

   (1) Li   (3) K
   (2) Na   (4) Rb   8_____

9. Which element is a liquid at room temperature (293 K)?

   (1) Br   (3) Ar
   (2) Cd   (4) K   9_____

10. Which element is a solid at STP and a good conductor of electricity?

    (1) iodine   (3) nickel
    (2) mercury   (4) sulfur   10_____

11. Which element is least dense?
    (1) oxygen   (3) sulfur
    (2) fluorine   (4) nitrogen   11_____

12. When cooled from room temperature, which substance would condense out first?

    (1) oxygen   (3) helium
    (2) hydrogen   (4) radon   12_____

13. Which element would float in water?

    (1) sodium   (3) cesium
    (2) magnesium   (4) radium   13_____

14. Which element has the greatest density at 298 K and 101.3 kPa?

    (1) calcium   (3) chlorine
    (2) carbon   (4) copper   14_____

1. 2 Open to Table S and locate bromine, atomic number 35. Read over to the Melting Point column. The melting point of bromine is 266 K. This would also be the freezing point of bromine.

2. 4 Standard temperature is 273 K (see Table A). Halogens are located in Group 17 of the Periodic Table. Using Table S, look up the melting points of these elements. Only I_2 has a melting point higher than 273 K, and therefore would be a solid at this temperature.

3. 3 Standard temperature is 273 K as shown in Table A. Using Table S, find the melting point for the given choices. Only Fe, with a melting point of 1,811 K, has a higher melting point than 273 K.

4. 2 All elements that have a boiling point below standard temperature (273 K) would be a gas at STP. The elements in Group 17 are the halogens. Table S shows that only F and Cl are gases at 273 K.

5. 3 The unknown element must have a melting point higher than 273 K to be a solid at this temperature. Choice 4, being element 17, is eliminated since it is a gas at 273 K. Element 12 and element 13 are metals. Metals are good conductors. Element 16, sulfur, is a non-metal and is a solid at this temperature. Solid non-metals exhibit properties of being brittle and poor conductors.

6. 1 As shown by Table S, most elements have melting points that are higher than 300 K. This would make most elements solids at this temperature.

7. 4 The Density column in Table S shows that gold (Au), with a density of 19.3 g/cm^3, has the greatest density of the given choices.

1. Which noble gas has the highest first ionization energy?

   (1) radon   (3) neon
   (2) krypton   (4) helium   1 _____

2. According to Reference Table S, which sequence correctly places the elements in order of increasing ionization energy?

   (1) H → Li → Na → K
   (2) I → Br → Cl → F
   (3) O → S → Se → Te
   (4) H → Be → Al → Ga   2 _____

3. As the elements of Group 1 on the Periodic Table are considered in order of increasing atomic radius, the ionization energy of each successive element generally

   (1) decreases
   (2) increases
   (3) remains the same   3 _____

4. Samples of four Group 15 elements, antimony, arsenic, bismuth, and phosphorus, are in the gaseous phase. An atom in the ground state of which element requires the least amount of energy to remove its most loosely held electron?

   (1) As   (3) P
   (2) Bi   (4) Sb   4 _____

5. What are two properties of most nonmetals?
   (1) high ionization energy and poor electrical conductivity
   (2) high ionization energy and good electrical conductivity
   (3) low ionization energy and poor electrical conductivity
   (4) low ionization energy and good electrical conductivity   5 _____

6. Which two characteristics are associated with metals?

   (1) low first ionization energy and low electronegativity
   (2) low first ionization energy and high electronegativity
   (3) high first ionization energy and low electronegativity
   (4) high first ionization energy and high electronegativity   6 _____

7. Which general trend is found in Period 2 on the Periodic Table as the elements are considered in order of increasing atomic number?

   (1) increasing first ionization energy
   (2) decreasing atomic mass
   (3) decreasing electronegativity
   (4) increasing atomic radius   7 _____

8. Explain, in terms of atomic structure, why cesium has a lower first ionization energy than rubidium.

_____

_____

9. From which of these atoms in the ground state can a valence electron be removed using the *least* amount of energy?

   (1) nitrogen   (3) oxygen
   (2) carbon   (4) chlorine   9 _____

10. Which element in Group 1 has the greatest tendency to lose an electron?

    (1) cesium   (3) potassium
    (2) rubidium   (4) sodium   10 _____

11. As the elements of Group 3 are considered in order from top to bottom, the first ionization energy of each successive element will

    (1) decrease
    (2) increase
    (3) remain the same   11 _____

12. An element that has a low first ionization energy and good conductivity of heat and electricity is classified as a

    (1) metal   (3) nonmetal
    (2) metalloid   (4) noble gas   12 _____

13. Which two characteristics are associated with nonmetals?
    (1) low first ionization energy and low electronegativity
    (2) low first ionization energy and high electronegativity
    (3) high first ionization energy and low electronegativity
    (4) high first ionization energy and high electronegativity   13 _____

14. Which electron dot symbol represents the atom in Period 4 with the highest first ionization energy?

    (1) $\ddot{X}$ (3) $\cdot\ddot{X}\colon$

    (2) $\ddot{X}\cdot$ (4) $\colon\!\ddot{X}\!\colon$ 14 _____

15. Which element in the halogen family has the highest first ionization energy?

    (1) carbon   (3) silcon
    (2) chlorine   (4) iodine   15 _____

16. Explain why helium and neon have such high ionization energies?

_____

_____

_____

Properties of Selected Elements

Table S – Properties of Selected Elements
First Ionization Energy
Answers
Set 1

1. **4**   Table S gives the first ionization energy value for helium as 2,372 kJ/mol. This is the highest first ionization of the given choices.

2. **2**   As given in the First Ionization Energy column in Table S, the elements in choice 2 are arranged in order of increasing ionization energy.

3. **1**   As one travels down a group of elements in the Periodic Table, the number of energy levels increases, causing the atomic radius to also increase. Open to Table S and compare the first ionization energy of the elements in Group 1. As one moves down this group, their radius increases and their first ionization energy decreases.

4. **2**   Table S gives the first ionization energy value for bismuth as 703 kJ/mol. This is the lowest value of the given elements. Bi would give up its outermost electron before the other 3 given elements.

5. **1**   As shown in Table S, nonmetals have high first ionization energies compared to metals. Nonmetals are also poor electrical conductors.

6. **1**   In Table S, it shows that the metals have low ionization energies compared to nonmetals. The same chart also shows that metals have low electronegativity values when compared to nonmetals.

7. **1**   Using Table S and the Periodic Table, look up the first ionization energy of each element in Period 2, starting with Li and ending with Ne. It shows that as the atomic number increases, the first ionization energy also increases.

8.   Answer: As atomic radius increases, valence electrons are more easily removed.

 or The force of attraction between the nucleus and the valence electrons decreases going down the group.

 or cesium has more shells, easier to remove electrons

 Explanation: The farther the valence electrons are from the nucleus, the lower the energy needed to remove the most loosely bound (outermost) valence electron.

1. Electronegativity is a measure of an atom's ability to

   (1) attract the electrons in the bond between the atom and another atom
   (2) repel the electrons in the bond between the atom and another atom
   (3) attract the protons of another atom
   (4) repel the protons of another atom

   1 _____

2. Based on electronegativity values, which type of elements tends to have the greatest attraction for electrons in a bond?

   (1) metals   (3) nonmetals
   (2) metalloids   (4) noble gases

   2 _____

3. Which element has atoms with the greatest attraction for electrons in a chemical bond?

   (1) beryllium   (3) lithium
   (2) fluorine   (4) oxygen

   3 _____

4. The ability of carbon to attract electrons is
   (1) greater than that of nitrogen, but less than that of oxygen
   (2) less than that of nitrogen, but greater than that of oxygen
   (3) greater than that of nitrogen and oxygen
   (4) less than that of nitrogen and oxygen

   4 _____

5. If the electronegativity difference between the elements in compound NaX is 2.1, what is element X?

   (1) bromine   (3) fluorine
   (2) chlorine   (4) oxygen

   5 _____

6. Which substance is correctly paired with its type of bonding?

   (1) NaBr—nonpolar covalent
   (2) HCl—nonpolar covalent
   (3) NH_3—polar covalent
   (4) Br_2—polar covalent

   6 _____

7. Which compound contains ionic bonds?

   (1) NO   (3) CaO
   (2) NO_2 (4) CO_2

   7 _____

8. Which substance contains a bond with the greatest ionic character?

   (1) KCl   (3) Cl_2
   (2) HCl   (4) F_2

   8 _____

9. Which of these formulas contains the most polar bond?

   (1) H-Br   (3) H-F
   (2) H-Cl   (4) H-I

   9 _____

10. Which substance contains nonpolar covalent bonds?

   (1) H_2 (3) $Ca(OH)_2$
   (2) H_2O (4) CaO

   10 _____

Base your answer to question 11 using the accompanying balanced equation.

$$2Na(s) + Cl_2(g) \rightarrow 2NaCl(s)$$

11. Explain, in terms of electronegativity, why the bonding in NaCl is ionic.

_____

12. Explain, in terms of electronegativity, why the H–F bond is expected to be more polar than the H–I bond.

_____

Base your answers to questions 13 through 15 using your knowledge of chemistry and the table below, which shows the electronegativity of selected elements of Period 2 of the Periodic Table.

13. On the grid below, set up a scale for electronegativity on the *y*-axis. Plot the data by drawing a best-fit line.

14. Using the graph, predict the electronegativity of nitrogen.

_____

Element	Atomic Number	Electronegativity
Beryllium	4	1.6
Boron	5	2.0
Carbon	6	2.6
Fluorine	9	4.0
Lithium	3	1.0
Oxygen	8	3.4

15. For these elements, state the trend in electronegativity in terms of atomic number.

_____

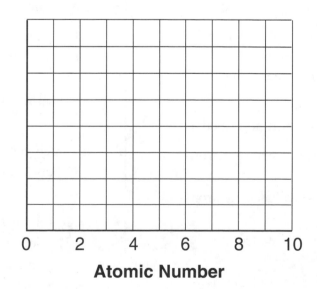

Properties of Selected Elements
Page 165

16. The strength of an atom's attraction for the electrons in a chemical bond is the atom's

    (1) electronegativity
    (2) ionization energy
    (3) heat of reaction
    (4) heat of formation   16 ____

17. Based on Reference Table S, the atoms of which of these elements have the strongest attraction for electrons in a chemical bond?

    (1) N   (3) P
    (2) Na   (4) Pt   17 ____

18. The bonds between hydrogen and oxygen in a water molecule are classified as

    (1) polar covalent
    (2) nonpolar covalent
    (3) ionic
    (4) metallic   18 ____

19. Compared to atoms of metals, atoms of nonmetals generally

    (1) have higher electronegativities
    (2) have lower first ionization energies
    (3) conduct electricity more readily
    (4) have lower electronegativities   19 ____

20. The bonds in BaO are best described as

    (1) covalent, because valence electrons are shared
    (2) covalent, because valence electrons are transferred
    (3) ionic, because valence electrons are shared
    (4) ionic, because valence electrons are transferred   20 ____

21. Which trends appear as the elements in Period 3 are considered from left to right?

    (1) Metallic character decreases, and electronegativity decreases.
    (2) Metallic character decreases, and electronegativity increases.
    (3) Metallic character increases, and electronegativity decreases.
    (4) Metallic character increases, and electronegativity increases.   21 ____

22. The degree of polarity of a chemical bond in a molecule of a compound can be predicted by determining the difference in the

    (1) melting points of the elements in the compound
    (2) densities of the elements in the compound
    (3) electronegativities of the bonded atoms in a molecule of the compound
    (4) atomic masses of the bonded atoms in a molecule of the compound   22 ____

23. Which elements combine by forming an ionic bond?

    (1) sodium and potassium
    (2) sodium and oxygen
    (3) carbon and oxygen
    (4) carbon and sulfur   23 ____

24. Which general trend is demonstrated by the Group 17 elements as they are considered in order from top to bottom on the Periodic Table?

    (1) a increase in electronegativity
    (2) a decrease in electronegativity
    (3) an increase in first ionization energy
    (4) an increase in nonmetallic behavior   24 ____

25. Which formula represents an ionic compound?

    (1) NaCl   (3) HCl

    (2) N_2O (4) H_2O 25 _____

26. In which molecule would the attraction between H and the atom to which it is bonded be the greatest?

    (1) HCl   (3) HBr

    (2) HF   (4) HI   26 _____

27. Which pair of atoms is held together by a covalent bond?

    (1) HCl   (3) NaCl

    (2) LiCl   (4) KCl   27 _____

28. The bond between Br atoms in a Br_2 molecule is

    (1) ionic and is formed by the sharing of two valence electrons

    (2) ionic and is formed by the transfer of two valence electrons

    (3) covalent and is formed by the sharing of two valence electrons

    (4) covalent and is formed by the transfer of two valence electrons

    28 _____

29. Which formula represents a nonpolar molecule containing polar covalent bonds?

    (1) H_2O (3) NH_3

    (2) CCl_4 (4) H_2 29 _____

30. Explain, in terms of electronegativity, why a P-Cl bond in a molecule of PCl_5 is more polar than a P-S bond in a molecule of P_2S_5.

_____

_____

Base your answer to question 31 using your knowledge of chemical bonding and the Lewis electron-dot diagrams of CO_2, and F_2 below.

$$:O::C::O: \qquad :F:F:$$

31. Explain, in terms of electronegativity, why a C–O bond in CO_2 is more polar than the F–F bond in F_2.

_____

_____

1. 1 By definition, electronegativity is the measure of the relative ability of that atom to attract bonding electrons to itself. The greater the electronegativity value, the greater the attraction.

2. 3 The electronegativity values, as shown in Table S, are highest for nonmetals. They would have the greatest attraction for electrons in a chemical bond compared to the other choices.

3. 2 Electronegativity is a value that measures the relative attraction the atom has for electrons in a chemical bond. The higher the electronegativity, the greater the attraction. Fluorine has the highest electronegativity (4.0) of the given elements.

4. 4 The lower the electronegativity, the less attraction that atom will have for electrons in a chemical bond. Carbon has an electronegativity value of 2.6. The electronegativity of nitrogen and oxygen are 3.0 and 3.4 respectively. Therefore carbon would have less attraction for electrons than nitrogen or oxygen.

5. 1 Na (sodium) has an electronegativity value of 0.9 and Br (bromine) has an electronegativity value of 3.0. The difference of these two values is 2.1.

6. 3 The difference in electronegativity between two elements is used to determine the type of bond formed between those elements.
    - If the difference in electronegativity (ΔEN) is less than 1.7, the bond is a covalent bond.
    - If the ΔEN is between 0 and 1.7, the bond is a polar covalent bond.
    - If the ΔEN is 0, the bond is a nonpolar covalent bond.
    - If the ΔEN is greater than 1.7, the bond is an ionic or electrovalent bond.

    Using the above rules, only choice 3 is correct. The difference between the electronegativity value of N and H is 0.8. This compound would contain polar covalent bonds.

7. 3 When there is a transfer of electrons between a metal and a nonmetal, an ionic compound is produced. Ionic bonds form when the difference between the electronegativity values of the two atoms is greater 1.7. The electronegativity for Ca and O is 1.0 and 3.4 respectively, giving a difference of 2.4. Therefore, the bond is ionic.

8. 1 The greater the difference in electronegativity, the greater the ionic character of the bond. KCl has the greatest electronegativity difference (2.4) of the given choices, thus would have the greatest ionic character. Cs is element 55 and Rb is element 37. From the Periodic Table, it shows that the Cs valence electron is located one further energy level out than Rb valence electron.

9. 3   The polarity of a bond can be determined from the difference between the electronegativity values for the two atoms. The larger the electronegativity difference, the more polar the bond. Referring to Table S, hydrogen has an electronegativity value of 2.2 and fluorine is 4.0, for a difference of 1.8. This is the largest difference of all given choices, making the H-F bond the most polar.

10. 1   If the difference between the electronegativities is 0, the resulting bond will be a nonpolar covalent bond. In H_2, the difference in electronegativities is 0 (2.2 – 2.2). Therefore, the bond is a nonpolar covalent bond.

11.   Ionic bonds form when the difference between the electronegativity values of the two atoms is greater than 1.7. The electronegativity for Na and Cl is 0.9 and 3.2, respectively. This is a difference of 2.3 and the bond is classified as ionic.

12.   The polarity of a bond can be determined from the difference between the electronegativity values of the two atoms. The larger the electronegativity difference, the more polar the bond. The electronegativity difference for H and F is 1.8, and the electronegativity difference for H-I is 0.5. Therefore, the H-F bond would be more polar than the H-I bond.

13.

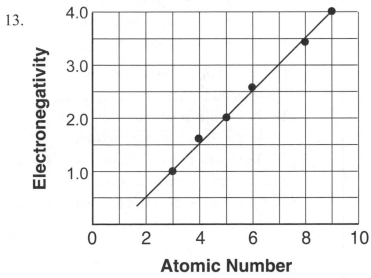

Explanation: Use an interval of 0.5, starting with 0 at the origin.

14.   Answer: The electronegativity is 3.0 (± 0.2)

Explanation: Nitrogen has an atomic number of 7. Locate this number on the atomic number axis and move upward until it intersects the graph line. Read over to the electronegativity axis to find a value of 3.0 for nitrogen.

15.   Answer: As the atomic number increases, the electronegativity increases.

Explanation: This relationship, as shown by the increasing slope of the graph, shows that as the atomic number increases, the electronegativity increases (a direct relationship).

---

1. An atom of which element has the largest atomic radius?

   (1) Fe   (3) Si
   (2) Mg   (4) Zn   1 _____

2. Which grouping of circles, when considered in order from the top to the bottom, best represents the relative size of the atoms of Li, Na, K, and Rb, respectively?

       (1)   (2)   (3)   (4)   2 _____

3. Which list of elements is arranged in order of increasing atomic radii?

   (1) Li, Be, B, C
   (2) Sr, Ca, Mg, Be
   (3) Sc, Ti, V, Cr
   (4) F, Cl, Br, I   3 _____

4. When an atom becomes a positive ion, the radius of the atom

   (1) decreases
   (2) increases
   (3) remains the same   4 _____

5. Which ion has the largest radius?

   (1) Na^+ (3) K^+
   (2) Mg^{2+} (4) Ca^{2+} 5 _____

6. What occurs when an atom loses an electron?

   (1) The atom's radius decreases and the atom becomes a negative ion.
   (2) The atom's radius decreases and the atom becomes a positive ion.
   (3) The atom's radius increases and the atom becomes a negative ion.
   (4) The atom's radius increases and the atom becomes a positive ion.   6 _____

7. As the elements in Group 17 on the Periodic Table are considered from top to bottom, what happens to the atomic radius and the metallic character of each successive element?

   (1) The atomic radius and the metallic character both increase.
   (2) The atomic radius increases and the metallic character decreases.
   (3) The atomic radius decreases and the metallic character increases.
   (4) The atomic radius and the metallic character both decrease.   7 _____

8. Compared to the radius of a chlorine atom, the radius of a chloride ion is

   (1) larger because chlorine loses an electron
   (2) larger because chlorine gains an electron
   (3) smaller because chlorine loses an electron
   (4) smaller because chlorine gains an electron

                        8 _____

      Properties of Selected Elements

Base your answer to question 9 using the information below and your knowledge of chemistry.

A metal, M, was obtained from a compound in a rock sample. Experiments have determined that this element is a member of Group 2 on the Periodic Table of the Elements.

9. Explain why the radius of a positive ion of element M is smaller than the radius of an atom of element M.

_____

_____

Base your answer to question 10 using the information below and on your knowledge of chemistry.

Potassium ions are essential to human health. The movement of dissolved potassium ions, K^+(aq), in and out of a nerve cell allows that cell to transmit an electrical impulse.

10. Explain, in terms of atomic structure, why a potassium ion is smaller than a potassium atom.

_____

_____

11. Explain, in terms of atomic structure, why the atomic radius of iodine is greater than the atomic radius of fluorine.

_____

_____

12. One electron is removed from both an Na atom and a K atom, producing two ions. Using principles of atomic structure, explain why the Na ion is much smaller than the K ion. Discuss both ions in your answer.

_____

_____

13. Fr - francium has the largest atomic radius of all given elements in Table S. Using the Periodic Table, explain why this is to be expected.

_____

_____

_____

14. Within Period 2 of the Periodic Table, as the atomic number increases, the atomic radius generally

(1) decreases
(2) increases
(3) remains the same   14 _____

15. Which list of elements from Group 2 on the Periodic Table is arranged in order of increasing atomic radius?

(1) Be, Mg, Ca   (3) Ba, Ra, Sr
(2) Ca, Mg, Be   (4) Sr, Ra, Ba   15 _____

16. Which of the following ions has the smallest radius?

(1) F$^-$ (3) K$^+$
(2) Cl$^-$ (4) Ca^{2+} 16 _____

17. As atomic number increases within Group 15 on the Periodic Table, atomic radius

(1) decreases, only
(2) increases, only
(3) decreases, then increases
(4) increases, then decreases   17 _____

18. When an atom loses one or more electrons, this atom becomes a

(1) positive ion with a radius smaller than the radius of this atom
(2) positive ion with a radius larger than the radius of this atom
(3) negative ion with a radius smaller than the radius of this atom
(4) negative ion with a radius larger than the radius of this atom   18 _____

19. An ion of which element has a larger radius than an atom of the same element?

(1) aluminum   (3) magnesium
(2) chlorine   (4) sodium   19 _____

20. Compared to the nonmetals in Period 2, the metals in Period 2 generally have larger

(1) ionization energies   (3) atomic radii
(2) electronegativities   (4) atomic numbers
                                    20 _____

21. Which trends are observed when the elements in Period 3 on the Periodic Table are considered in order of increasing atomic number?

(1) The atomic radius decreases, and the first ionization energy generally increases.
(2) The atomic radius decreases, and the first ionization energy generally decreases.
(3) The atomic radius increases, and the first ionization energy generally increases.
(4) The atomic radius increases, and the first ionization energy generally decreases.
                                    21 _____

22. Which characteristics both generally decrease when the elements in Period 3 on the Periodic Table are considered in order from left to right?

(1) nonmetallic properties and atomic radius
(2) nonmetallic properties and ionization energy
(3) metallic properties and atomic radius
(4) metallic properties and ionization energy
                                    22 _____

Properties of Selected Elements

Base your answers to question 23 on the information below.

The ionic radii of some Group 2 elements are given in the table below.

Ionic Radii of Some Group 2 Elements

Symbol	Atomic Number	Ionic Radius (pm)
Be	4	44
Mg	12	66
Ca	20	99
Ba	56	134

23. *a)* On the grid, mark an appropriate scale on the axis labeled "Ionic Radius (pm)."

b) On the same grid, plot the data from the data table. Circle and connect the points.

c) Estimate the ionic radius of strontium.

_____

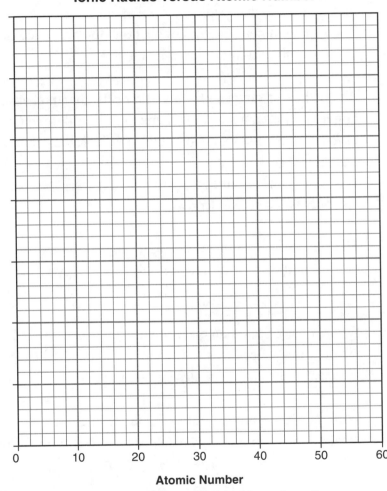

Ionic Radius Versus Atomic Number

Ionic Radius (pm)

Atomic Number

d) State the trend in ionic radius as the elements in Group 2 are considered in order of increasing atomic number.

_____

_____

e) Explain, in terms of electrons, why the ionic radius of a Group 2 element is smaller than its atomic radius.

_____

_____

Table S – Properties of Selected Elements
Atomic Radius
Answers
Set 1

1. 2   Table S gives the atomic radius of Mg as 129 pm. This value is the greatest of the given choices.

2. 1   When they speak of relative sizes of an atom, they are referring to the atomic radius. As one moves down in a group, the atomic radius increases due to more energy levels in the atom. The given elements are in Group 1 and their atomic radii increases from Li to Rb. This is shown in the Atomic Radius column in Table S.

3. 4   As one moves down a group of elements, the number of energy levels increases, producing a larger atomic radius. All elements in choice 4 are in Group 17. The increase in the atomic radius of these elements is shown in the Atomic Radius column of Table S.

4. 1   When an atom becomes a positive ion, it loses the electron(s) in the valence level or shell. When this occurs, its radius decreases.

5. 3   The radius of the K^+ and Ca^{2+} ions, having three energy levels each (2-8-8), will be larger than the Na^+ and Mg^{2+} ions, having only two energy levels (2-8). However, since the K atom is larger than the Ca atom (see Table S), the K^+ ion will be larger than the Ca^{2+} ion.

6. 2   When an atom loses an electron, its radius decreases. If this occurs to a Group 1 element it loses the outermost electron shell or valence shell. This causes the radius to decrease. When an atom loses an electron, it becomes a positive ion.

7. 1   With increasing atomic number in a given group, the atomic radius generally increases due to a greater number of energy levels. Also, as one moves down a group, the metallic properties also increase.

8. 2   Chlorine, being an active nonmetal, will gain an electron, becoming a negative ion. This gain of an electron causes the radius to increase.

9.   Answer: The ionic radius is smaller because the atom loses two electrons.
   or   The ion has one less occupied energy level.

   Explanation: Element M, being in Group 2, will lose the two electrons in its valence shell. The decrease in the number of electron levels causes a decrease in the radius.

10.   Answer: The potassium atom has an electronic configuration of 2-8-8-1 and the potassium ion has an electronic configuration of 2-8-8.

      or   The potassium ion (K^+) has only 3 energy levels.

      Explanation: As shown in the Periodic Table, the potassium atom contains 4 energy levels (shells). When it loses an electron, it now has only 3 energy levels, causing a decrease in the atomic radius.

11.   Answer: Iodine has more energy levels and its atomic radius would be larger than fluorine.

      or   The more energy levels an atom has, the larger its atomic radius will be.

      Explanation: With increasing atomic number in a given group, the number of electron levels in the atom increases and the atomic radius increases.

12.   Answer and explanation: The Periodic Table shows that sodium contains 3 energy levels. When it loses an electron, it will have only 2 energy levels. A potassium atom has 4 energy levels. When potassium loses an electron, it will contain 3 energy levels. Therefore, the sodium ion will have a smaller radius than the potassium ion.

13.   Answer: Fr is located in Period 7 and would have 7 energy levels causing it to have a large atomic radius.

      or   Fr is located in Group 1 and Period 7. This position with 7 energy levels and being in Group 1 gives Fr a large atomic radius.

      Explanation: The more energy levels an atom has the larger the atomic radius will be. Remember, with increasing atomic number in a given period, the atomic radius generally decreases. Thus, Group 1 elements usually will have a larger atomic radius than other groups in the same period.

Density	$d = \dfrac{m}{V}$	d = density m = mass V = volume
Mole Calculations	number of moles $= \dfrac{\text{given mass}}{\text{gram-formula mass}}$	
Percent Error	% error $= \dfrac{\text{measured value} - \text{accepted value}}{\text{accepted value}} \times 100$	
Percent Composition	% composition by mass $= \dfrac{\text{mass of part}}{\text{mass of whole}} \times 100$	
Concentration	parts per million $= \dfrac{\text{mass of solute}}{\text{mass of solution}} \times 1\,000\,000$	
	molarity $= \dfrac{\text{moles of solute}}{\text{liter of solution}}$	
Combined Gas Law	$\dfrac{P_1 V_1}{T_1} = \dfrac{P_2 V_2}{T_2}$	P = pressure V = volume T = temperature
Titration	$M_A V_A = M_B V_B$	M_A = molarity of H^+ M_B = molarity of OH^- V_A = volume of acid V_B = volume of base
Heat	$q = mC\Delta T$ $q = mH_f$ $q = mH_v$	q = heat H_f = heat of fusion m = mass H_v = heat of vaporization C = specific heat capacity ΔT = change in temperature
Temperature	$K = {}^\circ C + 273$	K = kelvin ${}^\circ C$ = degree Celsius

$$d = \frac{m}{V}$$

d = density
m = mass
V = volume

Overview:

Density is defined as mass per unit volume and is calculated by dividing the mass of a given sample of a substance by its volume. It is usually expressed in units of g/cm³ for solids or g/mL for liquids. The density of gases is usually expressed in units of g/L.

Example:

A student used a balance and a graduated cylinder to collect the following data:

Sample mass	10.23 g
Volume of water	20.0 mL
Volume of water and sample	21.5 mL

Calculate the density of the sample. Show your work. Include the appropriate number of significant figures and proper units.

Solution: The density equation is $d = m/V$
         $m = 10.23$ g   V of sample $= 21.5$ mL $- 20.0$ mL $= 1.5$ mL

Substitution: $d = \frac{10.23 \text{ g}}{1.5 \text{ mL}} = 6.82$ g/mL

Answer: $d = 6.8$ g/mL – to the correct number of significant figures.

Additional Information:

- Pressure and temperature can affect the density value of a substance. This is especially true for gases.

- Normally, solids are denser than liquids, while gases are the least dense.

- The exception to the above statement is water. The solid phase is less dense than the liquid phase.

- Density is an identifying property of matter. By calculating the density of an object, it greatly assists in the identification of the object.

- Density can be used to determined molecular masses of gases, knowing that one mole of any gas occupies a volume of 22.4 liters at STP.

1. A 2.00-liter sample of a gas has a mass of 1.80 grams at STP. What is the density, in grams per liter, of this gas at STP?

   (1) 0.900   (3) 11.2
   (2) 1.80   (4) 22.4   1 _____

2. The table below shows mass and volume data for four samples of substances at 298 K and 1 atmosphere.

   Masses and Volumes of Four Samples

Sample	Mass (g)	Volume (mL)
A	30.	60.
B	40.	50.
C	45	90.
D	90.	120.

   Which two samples could consist of the same substance?

   (1) A and B   (3) B and C
   (2) A and C   (4) C and D   2 _____

3. At STP, a 7.435-gram sample of an element has a volume of 1.65 cubic centimeters. The sample is most likely

   (1) Ta   (3) Te
   (2) Tc   (4) Ti   3 _____

4. What is the mass of 26.70 cm^3 of sodium? (see Table S)

   (1) 26.70 g   (3) 0.04 g
   (2) 25.98 g   (4) 13.56 g   4 _____

5. What is the volume of a 16.24 g sample of magnesium in grams per cubic centimeters? (see Table S)

   (1) 1.74   (3) 0.19
   (2) 16.24   (4) 9.33   5 _____

6. A gas has a density of 1.25 g/L at STP. What is the mass of one mole of this gas?

   (1) 14 g   (3) 42 g
   (2) 28 g   (4) 56 g   6 _____

7. *a)* A sample of an element has a mass of 34.261 grams and a volume of 3.8 cubic centimeters. To which number of significant figures should the calculated density of the sample be expressed?

   (1) 5   (3) 3
   (2) 2   (4) 4   a _____

   b) Determine the sample element's density.

   _____ g/cm^3

8. A 1.00-mole sample of neon gas occupies a volume of 24.4 liters at 298 K and 101.3 kilopascals. In the space below, calculate the density of this sample. Your response must include both a correct numerical setup and the calculated result.

     Important Formulas and Equations

Base your answers to question 9 using the information below and your knowledge of chemistry.

9. A student used a balance and a graduated cylinder to collect the following data:

Sample mass	12.74 g
Volume of water	25.0 mL
Volume of water and sample	32.3 mL

a) Calculate the density of the element. Show your work. Include the appropriate number of significant figures and proper units.

b) What error is introduced if the volume of the sample is determined first?

Base your answers to question 10 using the information below and your knowledge of chemistry.

Archimedes (287–212 BC), a Greek inventor and mathematician, made several discoveries important to science today. According to a legend, Hiero, the king of Syracuse, commanded Archimedes to find out if the royal crown was made of gold, only. The king suspected that the crown consisted of a mixture of gold, tin, and copper.

Archimedes measured the mass of the crown and the total amount of water displaced by the crown when it was completely submerged. He repeated the procedure using individual samples, one of gold, one of tin, and one of copper. Archimedes was able to determine that the crown was not made entirely of gold without damaging it.

10. *a*) Identify one physical property that Archimedes used in his comparison of the metal samples.

_____

b) Determine the volume of a 75-gram sample of gold at STP.

11. The mass of an unknown solid is 10.04 grams and the volume is 8.21 cm³. What is the density to the correct significant figures?

    (1) 1.2 g/cm³   (3) 1.222 g/cm³

    (2) 1.22 g/cm³   (4) .82 g/cm³   11 ____

12. A 10.0-gram sample of which element has the smallest volume at STP?

    (1) aluminum   (3) titanium

    (2) magnesium   (4) zinc   12 ____

Base your answer to question 13 using the information below and your knowledge of chemistry.

The decomposition of sodium azide, $NaN_3(s)$, is used to inflate airbags. On impact, the $NaN_3(s)$ is ignited by an electrical spark, producing $N_2(g)$ and $Na(s)$. The $N_2(g)$ inflates the airbag.

13. An inflated airbag has a volume of 5.00×10^4 cm³ at STP. The density of $N_2(g)$ at STP is 0.00125 g/cm³. What is the total number of grams of $N_2(g)$ in the airbag?

Base your answer to question 14 using the table below and your knowledge of chemistry.

Physical Properties of Four Gases

Name of Gas	hydrogen	hydrogen chloride	hydrogen bromide	hydrogen iodide
Molecular Structure	H–H	H–Cl	H–Br	H–I
Boiling Point (K) at 1 Atm	20.	188	207	237
Density (g/L) at STP	0.0899	1.64	?	5.66

14. The volume of 1.00 mole of hydrogen bromide at STP is 22.4 liters. The gram-formula mass of hydrogen bromide is 80.9 grams per mole. What is the density of hydrogen bromide at STP?

Important Formulas and Equations

15. *a)* Using the accompanying chart, calculate the volume of a tin block that has a mass of 95.04 grams at STP. Your response must include *both* a numerical setup and the calculated result.

Densities of Group 14 Elements

Element	Density at STP (g/cm^3)
C	3.51
Si	2.33
Ge	5.32
Sn	7.31
Pb	11.35

b) The tin's block's temperature changed from STP to 384°C. How would this affect it's density value?

c) The volume of a sample of Pb is 23.57 cm^3. What is the samples mass? Your response must include *both* a numerical setup and the calculated result to the correct significant number.

Base your answer to question 16 using the information below.

Graphite and diamond are two crystalline arrangements for carbon. The crystal structure of graphite is organized in layers. The bonds between carbon atoms within each layer of graphite are strong. The bonds between carbon atoms that connect different layers of graphite are weak because the shared electrons in these bonds are loosely held by the carbon atoms. The crystal structure of diamond is a strong network of atoms in which all the shared electrons are strongly held by the carbon atoms. Graphite is an electrical conductor, but diamond is not. At 25°C, graphite has a density of 2.2 g/cm^3 and diamond has a density of 3.51 g/cm^3.

The element oxygen can exist as diatomic molecules, O_2, and as ozone, O_3. At standard pressure the boiling point of ozone is 161 K.

16. *a)* Calculate the volume, in cm^3, of a diamond at 25°C that has a mass of 0.200 gram. Your response must include *both* a correct numerical setup and the calculated result.

b) Determine the mass of a 23.6 cm^3 sample of graphite that is at 25°C. Your response must include *both* a correct numerical setup and the calculated results

1. 1 The variables are: V = 2.00 L, m = 1.80 g

   Equation: $d = \frac{m}{V}$ Substituting: d = 1.80 g/2.00 L   Solving: d = 0.900 g/L

2. 2 Under identical conditions, two samples of the same substance will have the same density. Both substance A and C have a density of 0.5 g/mL, thus they could consist of the same substance.

3. 4 One must solve for the density and then match it to the density of the correct element from Table S.

   Equation: $d = \frac{m}{V}$ Substituting: d = 7.435 g/1.65 cm³   Solving: d = 4.50 g/cm³

   Using Table S, this density value matches that of Ti – titanium.

4. 2 The density of sodium is 0.97 g/cm³ as given in Table S. The given volume is 26.70 cm³.
   Equation: $m = (d)(V)$ Substituting: m = (0.97 g/cm³)(26.70 cm³) Solving: m = 25.98 g

5. 4 The density of magnesium is 1.74 g/cm³ as given in Table S. The given mass is 16.24 g.

   Equation: $V = \frac{m}{d}$ Substituting: V = $\dfrac{16.24\ g}{1.74\ g/cm^3}$ Solving: V = 9.33 cm³

6. 2 At STP, one mole of any gas has a volume of 22.4 L. In the density equation, let m = the mass of one mole and V = 22.4 L. Substituting: 1.25 g/L = $\dfrac{m}{22.4\ L}$ Solving: m = (1.15 g/L)(22.4 L) = 28 g

7. a) 2 The volume measurement of 3.8 has 2 significant figures. The mass of 34.261 has 5 significant figures. In carrying out calculations, the general rule is that the accuracy of a calculated result is limited by the least accurate measurement involved in the calculation. Therefore, the calculated density should be expressed to 2 significant figures.

   b) Answer: 9.0 g/cm³

   Solving: d = 34.261 g/3.8 cm³

8.	Numerical setup:

$$d = \frac{m}{V} = \frac{20.180 \text{ g}}{24.4 \text{ L}} \quad or \quad \frac{20}{24.4}$$

Calculated result:

$d = 0.827$ g/L   or   0.83 g/L Significant figures do not need to be shown.

Explanation: This is a density problem involving the variables of mass and volume. The pressure and temperature values are not involved in this problem and can be ignored. The volume (V) is 24.4 liters. The mass (m) of one mole of neon is equal to its gram-atomic mass. The Periodic Table shows that neon has a gram-atomic mass of 20.180 g. Substitution and dividing gives you the correct answer as shown above.

9. *a)* Answer: 1.7 g/mL

Numerical setup: $\dfrac{12.74}{32.3 - 25.0}$ or $\dfrac{12.74}{7.3}$ or $\dfrac{12.74 \text{ g}}{7.3 \text{ mL}}$

Explanation: The volume of the sample is 7.3 mL (32.3 mL – 25.0 mL). The mass of the sample is 12.74 g. Equation: $d = \dfrac{m}{V}$

Substituting: $d = \dfrac{12.74 \text{ g}}{7.3 \text{ mL}}$ Solving: $d = 1.7$ g/mL

b) Answer: The density would increase because the sample was wet when weighed.

Explanation: If the volume is obtained by displacement of water first, the sample would be wet. This would cause the mass to increase. If mass increases and the volume remains the same, the density value will increase.

10. *a)* Answer: density *or* mass *or* volume

Explanation: By measuring the amount of displaced water, the physical property of volume was obtained. Mass is another physical property. Using the measurements of volume and mass, the density was calculated. The density of pure gold would be higher than the density of a mixture of gold, tin, and copper.

b) Answer: V = 3.88 cm^3 *or* 3.9 cm^3

Explanation: The density of gold (Au) from Table S is 19.3 g/cm^3. The given mass is 75 g.

Equation: $V = \dfrac{m}{d}$ Substituting: $V = \dfrac{75 \text{ g}}{19.3 \text{ g/cm}^3}$ Solving: V = 3.88 cm^3

$$\text{number of moles} = \frac{\text{given mass}}{\text{gram-formula mass}}$$

Overview:

The mole in chemistry is the Metric unit which is used as a measure of the amount of substance (see Table D). It is defined as 6.02×10^{23} (the Avogadro number) structural particles (atoms, ions, molecules, electrons, etc.). For a specific substance, one mole is a gram-formula mass of that substance, which is the sum of the gram atomic masses of the elements in the substance as indicated by the formula of the substance.

Example:

Determine the number of moles of CH_3Br in 19 grams of CH_3Br.

Solution: First obtain the gram-formula mass of CH_3Br. This is done by getting the sum of the gram atomic masses of each element from the Periodic Table.

From the Periodic Table:

$$C = 12 \text{ g}$$
$$3H = 3 \text{ g}$$
$$\underline{Br = 80 \text{ g}}$$
$$\text{gram-formula mass} = 95 \text{ grams/mol}$$
$$\text{given mass} = 19 \text{ grams}$$

Substitution: number of moles = $\dfrac{19 \text{ grams}}{95 \text{ grams/mol}}$

Answer: 0.20 mol

Additional Information:

- Two moles of a substance would have twice the gram formula mass and it would contain twice the Avogradro number (6.02×10^{23}) of particles.

- The gram formula mass of a gas equals the gas density at STP × 22.4 liters.

- At STP, the molar volume of any gas is 22.4 liters.

1. The gram formula mass of NH_4Cl is

   (1) 22.4 g/mole   (3) 53.5 g/mole
   (2) 28.0 g/mole   (4) 95.5 g/mole   1 _____

2. What is the gram-formula mass of $Ca_3(PO_4)_2$?

   (1) 248 g/mol   (3) 279 g/mol
   (2) 263 g/mol   (4) 310. g/mol   2 _____

3. What is the mass of 5.2 moles of NO_3?

   (1) 62 g   (3) 11.9 g
   (2) 322.4 g   (4) 161.2 g   3 _____

4. The number of moles in 24 grams of $MgBr_2$ is

   (1) $\frac{24}{104}$ mole   (3) $\frac{104}{184}$ mole

   (2) $\frac{24}{184}$ mole   (4) $\frac{184}{24}$ mole   4 _____

5. How many moles are there in 222.3 grams of $Ca(OH)_2$?

   (1) 1.5 mol   (3) 5.5 mol
   (2) 3.0 mol   (4) 22.3 mol   5 _____

6. Determine the number of moles of CH_3Br in 47.5 grams of CH_3Br.

7. What is the mass of 4.76 moles of Na_3PO_4?

Base your answer to question 8 on the information and equation below.

Antacids can be used to neutralize excess stomach acid. Brand A antacid contains the acid neutralizing agent magnesium hydroxide, $Mg(OH)_2$. It reacts with HCl(aq) in the stomach according to the following balanced equation:

$$2\ HCl(aq) + Mg(OH)_2(s) \rightarrow MgCl_2(aq) + 2\ H_2O(\ell)$$

8. In the space provided below, show a correct numerical setup for calculating the number of moles of $Mg(OH)_2$ in an 8.40-gram sample.

9. The molar mass of $Ba(OH)_2$ is

(1) 154.3 g   (3) 171.3 g
(2) 155.3 g   (4) 308.6 g   9 _____

10. The number of moles in a 12.0-gram sample of Cl_2 is

(1) $\frac{12.0}{35.5}$ mole (3) 12.0 moles

(2) $\frac{12.0}{71.0}$ mole (4) 12.0×35.5 moles

10 _____

11. How many moles are there in a sample of 49 grams of H_2SO_4?

(1) 0.5 mol   (3) 1.5 mol
(2) 1.0 mol   (4) 2.0 mol   11 _____

12. What is the mass of 3.45 moles of NH_4NO_3?

(1) 80 g   (3) 189 g
(2) 158 g   (4) 276 g   12 _____

13. The sum of the atomic masses of the atoms in one molecule of $C_3H_6Br_2$ is called the

(1) formula mass
(2) isotopic mass
(3) percent abundance
(4) percent composition   13 _____

14. A 1.0-mole sample of krypton gas has a mass of

(1) 19 g   (3) 39 g
(2) 36 g   (4) 84 g   14 _____

15. The gram-formula mass of NO_2 is defined as the mass of

(1) one mole of NO_2
(2) one molecule of NO_2
(3) two moles of NO
(4) two molecules of NO   15 _____

16. In the space, show a correct numerical setup for calculating the number of moles of CO_2 present in 11 grams of CO_2.

17. The decomposition of sodium azide, $NaN_3(s)$, is used to inflate airbags. On impact, the $NaN_3(s)$ is ignited by an electrical spark, producing $N_2(g)$ and $Na(s)$. The $N_2(g)$ inflates the airbag.

What is the number of moles present in a 52.0-gram sample of $NaN_3(s)$?

Set 1

1. 3 The gram atomic masses of nitrogen, hydrogen and chlorine are 14.0 g, 1.0 g, and 35.5 g respectively rounded to the nearest 10th. Using these masses, the gram-formula mass of NH_4Cl is 14 g + (4 × 1.0 g) + 35.5 g = 53.5 g/mol.

2. 4 The gram atomic masses of Ca, P, and O are 40 g, 31 g, and 16 g respectively. Using these atomic masses the gram formula mass of $Ca_3(PO_4)_2$ is (40 × 3) + (31 × 2) + (16 × 8) = 310. g/mol.

3. 2 The gram-formula mass of NO_3 is 62 g/mol (14 g + 16 g + 16 g + 16 g).
    Substituting into the Mole Calculations equation:
    $$5.2 \text{ moles} = \frac{\text{given mass}}{62 \text{ g/mol}}$$
    Solving: given mass = (5.2 mole)(62 g/mol) = 322.4 g

4. 2 Open to Table T and locate the Mole Calculations formula. The gram-formula mass of $MgBr_2$ is 184 g/mole (24 g + 80 g + 80 g). The given mass is 24 g.
    Substituting: number of moles = $\frac{24}{184}$ mole

5. 2 The gram-formula mass of $Ca(OH)_2$ is 74.1 g/mol (40.1 g + 16 g + 16 g + 1 g + 1 g).
    Substituting: number of moles = $\frac{222.3 \text{ g}}{74.1 \text{ g/mol}}$
    Solving: number of moles = 3.0 mol

6. Answer: 0.50 mol (significant figures do not need to be shown)

    Explanation: The gram-formula mass of CH_3Br is 95 g/mol (12 g + 3 g + 80 g). The given mass is 47.5 g.
    Substituting: number of moles = $\frac{47.5 \text{ g}}{95 \text{ g/mol}}$
    Solving: 0.50 mol

7. Answer: 781 g (significant figures do not need to be shown)

    Explanation: The gram-formula of Na_3PO_4 is 164 g/mol. From the given information:
    number of moles = 4.76, given mass = X
    Substituting: $4.76 \text{ mol} = \frac{X}{164 \text{ g/mol}}$

    Solving: X = (4.76 mol)(164 g/mol), mass of 4.76 moles of Na_3PO_4 = 781 g

8. Numerical setup: $8.40 \text{ g} \times \frac{1 \text{ mole}}{58.3 \text{ g}}$ or $\frac{8.4}{58.3 \text{ g}}$

    Explanation: The gram-formula mass of $Mg(OH)_2$ is 58.3 g/mol, the given mass is 8.40 g.
    Using the Mole Calculations equation: number of moles of $Mg(OH)_2$ = $\frac{8.40 \text{ g}}{58.3 \text{ g/mol}}$.

$$\% \text{ error} = \frac{\text{measured value} - \text{accepted value}}{\text{accepted value}} \times 100$$

Overview:

Percent error is a means of comparing two numbers to obtain an indication of how close the numbers correspond to each other. This is especially useful at the end of many lab exercises to determine the accuracy of the overall measurement-calculation process. The two numbers are the calculated or measured value and the accepted value or known value is obtained from a handbook or reference book.

Example:

A student intended to make a salt solution with a concentration of 8.0 grams of solute per liter of solution. When the student's solution was analyzed, it was found to contain 8.90 grams of solute per liter of solution. What was the percent error in the concentration of the solution?

> Solution: measured value = 8.90 grams
> accepted value = 8.0 grams
> measured value – accepted value = 8.90 g – 8.0 g = 0.9 g

> Substitution: $\% \text{ error} = \dfrac{0.9 \text{ grams}}{8.0 \text{ grams}} \times 100$

> Answer: 11.3% *or* 11%

Additional Information:

- The major mistake in this equation is to substitute the wrong value for accepted value. Be careful here.

- If accepted value is larger than measured value, you may subtract the measured value from the accepted value to obtain a positive answer.

1. A student calculated the percent by mass of water in a hydrate as 14.2%. A hydrate is a compound that contains water as part of its crystal structure. If the accepted value is 14.7%, the student's percent error was

    (1) $\frac{0.5}{14.2} \times 100$ (3) $\frac{0.5}{14.7} \times 100$

    (2) $\frac{14.7}{14.2} \times 100$ (4) $\frac{14.2}{14.7} \times 100$ 1 _____

2. A student determined in the laboratory that the percent by mass of water in $CuSO_4 \cdot 5H_2O$ is 40.0%. If the accepted value is 36%, what is the percent of error?

    (1) 0.11%   (3) 11%
    (2) 1.1%   (4) 4.0%   2 _____

3. A student intended to make a salt solution with a concentration of 10.0 grams of solute per liter of solution. When the student's solution was analyzed, it was found to contain 8.90 grams of solute per liter of solution. What was the percent error in the concentration of the solution?

    (1) 1.10%   (3) 11.0%
    (2) 8.90%   (4) 18.9%   3 _____

4. A student calculated the percent by mass of water in a sample of $BaCl_2 \cdot 2H_2O$ to be 16.4%, but the accepted value is 14.8%. What was the student's percent error?

    (1) $\frac{14.8}{16.4} \times 100\%$ (3) $\frac{1.6}{14.8} \times 100\%$

    (2) $\frac{16.4}{14.8} \times 100\%$ (4) $\frac{14.8}{1.6} \times 100\%$ 4 _____

5. A student determines the density of zinc to be 7.56 grams per milliliter. If the accepted density is 7.14 grams per milliliter, what is the student's percent error?

    *   Show a correct numerical setup.
    *   Record your answer.

6. In an experiment, the molarity of KOH(aq) was determined to be 0.95 M. The actual molarity was 0.83 M. What is the percent error in the trial?

7. A student measured the wavelength of hydrogen's visible red spectral line to be 627 nanometers. The accepted value is 656 nanometers. What is the student's percent error?

(1) 0.04%   (3) 4.4%

(2) 1.4%   (4) 4.6%   7_____

8. A student determines the density of an element to be 1.56 grams per cubic centimeter. If the accepted value is 1.68 grams per cubic centimeter, what is the student's percent error?

(1) 7.7%   (3) 7.1%

(2) 3.3%   (4) 5.9%   8 _____

9. A hydrated compound contains water molecules within its crystal structure. The percent composition by mass of water in the hydrated compound $CaSO_4 \cdot 2H_2O$ has an accepted value of 20.9%. A student did an experiment and determined that the percent composition by mass of water in $CaSO_4 \cdot 2H_2O$ was 21.4%.

In the space below, calculate the percent error of the student's experimental result. Your response must include both a correct numerical setup and the calculated result.

Base your answer to question 10 on the information below.

The accepted values for the atomic mass and percent natural abundance of each naturally occurring isotope of silicon are given in the accompanying data table.

Naturally Occurring Isotopes of Silicon

Isotope	Atomic Mass (atomic mass units)	Percent Natural Abundance (%)
Si-28	27.98	92.22
Si-29	28.98	4.69
Si-30	29.97	3.09

10. A scientist calculated the percent natural abundance of Si-30 in a sample to be 3.29%. Determine the percent error for this value.

Important Formulas and Equations

1. **3** Locate the Percent Error equation in Table T. The student's measured value (calculated value) is 14.2% and the acceptable value is 14.7%.

   Substituting: % error $= \dfrac{14.7\% - 14.2\%}{14.7\%} \times 100 = \dfrac{0.5}{14.7} \times 100$

2. **3** The measured value is 40% and the accepted value is 36%.

   Substituting: % error $= \dfrac{40\% - 36\%}{36\%} \times 100 = \dfrac{4}{36} \times 100$. Solving: % error $= 0.11 \times 100 = 11\%$

3. **3** The measured value is 8.90 g, while the accepted value is 10.0 g.

   Substituting: % error $= \dfrac{10.0 \text{ g} - 8.9 \text{ g}}{10.0 \text{ g}} \times 100$ Solving: % error $= .11 \times 100 = 11.0\%$

4. **3** The measured value is 16.4%, while the accepted value is 14.8%.

   Substituting: % error $= \dfrac{16.4\% - 14.8\%}{14.8\%} \times 100 = \dfrac{1.6}{14.8} \times 100$

5. Correct numerical setup: $\dfrac{7.56 \text{ g} - 7.14 \text{ g}}{7.14 \text{ g}} \times 100$

   Answer: 5.8% *or* 5.9% *or* 6% Significant figures are not needed.

   Explanation: Substituting into the percent error equation gives the correct numerical setup.

   Substitution: % error $= \dfrac{7.56 \text{ g} - 7.14 \text{ g}}{7.14 \text{ g}} \times 100$. Solving: $\dfrac{0.42}{7.14} \times 100 = 5.8\%$

6. Answer: 14% *or* 14.46% *or* 14.5%

   Explanation: The measured value of the molarity of KOH(aq) is 0.95 M, and the accepted value is 0.83 M.

   Substituting: % error $= \dfrac{0.95 \text{ M} - 0.83 \text{ M}}{0.83 \text{ M}} \times 100$ Solving: $\dfrac{0.12}{0.83} \times 100 = 14.5\%$

$$\text{\% composition by mass} = \frac{\text{mass of part}}{\text{mass of whole}} \times 100$$

Overview:

Percent composition is used to determine what part of a compound one element contributes to its makeup. The mass of the whole is the mass of one mole (gram formula mass) of the compound.

Example:

A hydrated salt is a solid that includes water molecules within its crystal structure. A student heated a 9.10 gram sample of hydrated salt to a constant mass of 5.14 grams. What percent by mass of water did the salt contain?

> Explanation: 9.10 g sample of hydrated salt was heated to remove the water. The salt had a mass of 5.14 g. Thus, the mass of the water was 3.96 g (9.10 g – 5.14 g).

> Solution: mass of part = (H_2O) = 3.96 g
> mass of the whole = 9.10 g

> Substitution: % composition by mass = $\dfrac{3.96 \text{ g}}{9.10 \text{ g}} \times 100 = 43.5\%$

Additional Information:

- If you know the percent composition of each element within a compound, you can use this to arrive at the compound's formula.

1. The percent by mass of hydrogen in NH_3 is equal to

   (1) $\frac{17}{1} \times 100$ (3) $\frac{1}{17} \times 100$

   (2) $\frac{17}{3} \times 100$ (4) $\frac{3}{17} \times 100$ 1 _____

2. The percent by mass of calcium in the compound calcium sulfate ($CaSO_4$) is approximately

   (1) 15%   (3) 34%
   (2) 29%   (4) 47%   2 _____

3. A sample of a substance containing only magnesium and chlorine was tested in the laboratory and was found to be composed of 74.5% chlorine by mass. If the total mass of the sample was 190.2 grams, what was the mass of the magnesium?

   (1) 24.3 g   (3) 70.9 g
   (2) 48.5 g   (4) 142 g   3 _____

4. What is the percent by mass of oxygen in propanal, CH_3CH_2CHO?

   (1) 10.0%   (3) 38.1%
   (2) 27.6%   (4) 62.1%   4 _____

5. In which compound is the percent composition by mass of chlorine equal to 42%?

   (1) $HClO$ (gram-formula mass = 52 g/mol)
   (2) $HClO_2$ (gram-formula mass = 68 g/mol)
   (3) $HClO_3$ (gram-formula mass = 84 g/mol)
   (4) $HClO_4$ (gram-formula mass = 100. g/mol)

   5 _____

6. A hydrate is a compound that includes water molecules within its crystal structure. During an experiment to determine the percent by mass of water in a hydrated crystal, a student found the mass of the hydrated crystal to be 4.10 grams. After heating to constant mass, the mass was 3.70 grams. What is the percent by mass of water in this crystal?

   (1) 90.%   (3) 9.8%
   (2) 11%   (4) 0.40%   6 _____

7. Determine the percent composition by mass of oxygen in the compound $C_6H_{12}O_6$.

   • Show a correct numerical setup.

   • Record your answer.

8. The percent composition by mass of magnesium in $MgBr_2$ (gram-formula mass = 184 grams/mole) is equal to

   (1) $\dfrac{24}{184} \times 100$

   (2) $\dfrac{17}{3} \times 100$

   (3) $\dfrac{184}{24} \times 100$

   (4) $\dfrac{184}{160.} \times 100$ 8 _____

9. The percent by mass of carbon in $HC_2H_3O_2$ is equal to

   (1) $\dfrac{12}{60} \times 100$

   (2) $\dfrac{24}{60} \times 100$

   (3) $\dfrac{60}{24} \times 100$

   (4) $\dfrac{60}{12} \times 100$ 9 _____

10. The percentage by mass of Br in the compound $AlBr_3$ is closest to

    (1) 10.%   (3) 75%
    (2) 25%   (4) 90.%   10 _____

11. What is the percent by mass of oxygen in H_2SO_4? [formula mass = 98]

    (1) 16%   (3) 65%
    (2) 33%   (4) 98%   11 _____

12. A hydrated salt is a solid that includes water molecules within its crystal structure. A student heated a 12.10-gram sample of a hydrated salt to a constant mass of 5.41 grams. What percent by mass of water did the salt contain?

    (1) 3.69%   (3) 40.5%
    (2) 16.8%   (4) 55.3%   12 _____

Base your answers to question 13 using the information below and your knowledge of chemistry.

Gypsum is a mineral that is used in the construction industry to make drywall (sheetrock). The chemical formula for this hydrated compound is $CaSO_4 \cdot 2H_2O$. A hydrated compound contains water molecules within its crystalline structure. Gypsum contains 2 moles of water for each 1 mole of calcium sulfate.

13. *a)* What is the gram formula mass of $CaSO_4 \cdot 2H_2O$?

    b) Show a correct numerical setup for calculating the percent composition by mass of water in this compound.

    c) Record your answer.

14. At STP, iodine, I_2, is a crystal, and fluorine, F_2, is a gas. Iodine is soluble in ethanol, forming a tincture of iodine. A typical tincture of iodine is 2% iodine by mass.

Determine the total mass of I_2 in 25 grams of this typical tincture of iodine.

Base your answers to question 15 on the information below.

Litharge, PbO, is an ore that can be roasted (heated) in the presence of carbon monoxide, CO, to produce elemental lead. The reaction that takes place during this roasting process is represented by the balanced equation below.

$$PbO(s) + CO(g) \rightarrow Pb(\ell) + CO_2(g)$$

15 a) Calculate the percent composition by mass of oxygen in litharge. Your response must include both a numerical setup and the calculated result.

b) Determine the oxidation number of carbon in carbon monoxide. _____

Base your answers to question 16 on the information below.

Hydrogen peroxide, H_2O_2, is a water-soluble compound. The concentration of an aqueous hydrogen peroxide solution that is 3% by mass H_2O_2 is used as an antiseptic. When the solution is poured on a small cut in the skin, H_2O_2 reacts according to the balanced equation below.

$$2H_2O_2 \rightarrow 2H_2O + O_2$$

16. a) Calculate the total mass of H_2O_2 in 20.0 grams of an aqueous H_2O_2 solution that is used as an antiseptic. Your response must include both a numerical setup and the calculated result.

b) Determine the gram-formula mass of H_2O_2.

c) Identify the type of chemical reaction represented by the balanced equation.

1. 4 Referring to the Periodic Table, the gram-formula mass of NH_3 is 17 g/mol (14 g + 3 g). The percent of composition of H = $\dfrac{\text{mass of hydrogen}}{\text{mass of the whole}} \times 100$.

   Substituting: % composition by mass of hydrogen = $\dfrac{3}{17} \times 100$

2. 2 Referring to the Periodic Table, the gram-formula mass for $CaSO_4$ (mass of whole) is 136 g/mol (Ca = 1 × 40 g, S = 1 × 32 g and O = 4 × 16 g = 64 g). The mass of the part (calcium) is 40 g.

   Substituting and solving: % composition by mass of calcium = $\dfrac{40}{136} \times 100 = 29\%$

3. 2 The substance is 74.5% Cl by mass. Therefore the substance must be 25.5% Mg (100% – 74.5%). The mass of the whole is 190.2 g.

   Substituting: $25.5\% = \dfrac{\text{mass of Mg}}{190.2 \text{ g}} \times 100 \quad \text{mass of Mg} = \dfrac{(25.5\%)(190.2 \text{ g})}{100\%}$

   Solving: mass of Mg = $\dfrac{4850.1}{100} = 48.5$ g

4. 2 Referring to the Periodic Table, the gram-formula mass for CH_3CH_2CHO is 58 g/mol (12 g + (3 × 1 g) + 12 g + (2 × 1 g) + 12 g + 1 g + 16 g). The mass of the part (oxygen) is 16 g.

   Substituting and solving: % composition by mass of oxygen = $\dfrac{16 \text{ g}}{58 \text{ g}} \times 100 = 27.6\%$

5. 3 The % composition by mass of Cl is 42%. The mass of part (Cl) in each compound is 35.5 g (see Periodic Table).

   Substituting: $42\% = \dfrac{35.5 \text{ g}}{\text{mass of whole}} \times 100$

   Solving: mass of whole = $\dfrac{35.5 \text{ g}}{42} \times 100 = 84$ g/mol

6. 3 The mass of water is 0.40 g (4.10 g – 3.70 g) and the mass of the whole is 4.10 g.

   Substituting and solving: % composition by mass of water = $\dfrac{0.40 \text{ g}}{4.10 \text{ g}} \times 100 = 9.8\%$

7. Numerical setup: % composition by mass of O = $\dfrac{96}{180} \times 100$ (units may be shown)

   Answer: 53.3% *or* 53%   Significant numbers do not need to be shown.

   Explanation: From the Periodic Table, the gram-formula mass of $C_6H_{12}O_6$ is 180 g/mol. The mass of oxygen is 96 g.

   Substituting and solving: % composition by mass of O = $\dfrac{96 \text{ g}}{180 \text{ g}} \times 100 = 53.3\%$

$$\text{parts per million} = \frac{\text{grams of solute}}{\text{grams of solution}} \times 1\,000\,000$$

$$\text{molarity} = \frac{\text{moles of solute}}{\text{liters of solution}}$$

Overview:

The expressions given are used to indicate the concentration or strength of a solution. A solution consists of two components: the solute, or dissolved substance, and the solvent, or the dissolving substance. The most common solvent is water. In the parts per million equation, the grams of solution is the total mass of the solute and the solvent.

Example – parts per million:

An aqueous solution has 0.04 gram of oxygen dissolved in 1000. grams of water. Calculate the dissolved oxygen concentration of this solution in parts per million.

Equation: $\text{parts per million} = \frac{\text{grams of solute}}{\text{grams of solution}} \times 1\,000\,000$

Solution: grams of solute = 0.04 gram

grams of solution = 1000. grams of water + 0.04 grams of oxygen = 1000.04 grams

Substitution: $\text{parts per million} = \frac{0.04 \text{ gram}}{1000.04 \text{ grams}} \times 1\,000\,000$

Answer: Concentration = 40. ppm

Example – molarity:

What is the molarity of a solution that contains 0.50 mole of NaOH in 0.50 liter of solution?

Equation: $\text{molarity} = \frac{\text{moles of solute}}{\text{liters of solution}}$

Substitution: $\text{molarity} = \frac{0.50 \text{ mole of NaOH}}{0.50 \text{ liter of solution}}$

Answer: Concentration = 1.0 M

Additional Information:

- If the molarity of a solution is known, the number of moles of solute in a given volume of that solution can be found using the equation: moles of solute = molarity × liters of solution.

1. What is the concentration of a solution, in parts per million, if 0.02 gram of Na_3PO_4 is dissolved in 1000 grams of water?

   (1) 20 ppm   (3) 0.2 ppm
   (2) 2 ppm   (4) 0.02 ppm   1 _____

2. What is the concentration of a solution in parts per million, if 0.089 gram of NaCl is dissolved in 250 grams of water?

   (1) 0.03 ppm   (3) 35.6 ppm
   (2) 0.27 ppm   (4) 356 ppm   2 _____

3. An aqueous solution has 0.0070 gram of oxygen dissolved in 1000. grams of water. Calculate the dissolved oxygen concentration of this solution in parts per million. Your response must include both a correct numerical setup and the calculated result.

Base your answer to question 4 using the information below and your knowledge of chemistry.

A town located downstream from a chemical plant was concerned about fluoride ions from the plant leaking into its drinking water. According to the Environmental Protection Agency, the fluoride ion concentration in drinking water cannot exceed 4 ppm. The town hired a chemist to analyze its water. The chemist determined that a 175-gram sample of the town's water contains 0.000250 gram of fluoride ions.

4. How many parts per million of fluoride ions are present in the analyzed sample?

5. More than 60 parts per million of dissolved minerals is considered to be hard water. A 750-gram sample of water was found to contain 0.04 gram of minerals. Would this sample of water be considered hard water? Explain your answer.

_____

_____

Important Formulas and Equations

6. If 0.025 gram of $Pb(NO_3)_2$ is dissolved in 100. grams of H_2O, what is the concentration of the resulting solution, in parts per million?

   (1) 2.5×10^{-4} ppm
   (2) 2.5 ppm
   (3) 250 ppm
   (4) 4.0×10^3 ppm

   6 _____

7. What is the concentration of a solution in parts per million, if 0.45 gram of KNO_3 is dissolved in 1000. grams of water?

   (1) 450 ppm
   (2) 4.5×10^{-5} ppm
   (3) 4.5×10^{-6} ppm
   (4) 225 ppm

   7 _____

8. An aqueous solution contains 300. parts per million of KOH. Determine the number of grams of KOH present in 1000. grams of this solution.

$$300 = \frac{x}{1000} \times \frac{1,000,000}{1} \qquad \frac{300}{1} = \frac{1,000}{1}$$

9. An aqueous solution has 0.0023 gram of CO_2 dissolved in 500. grams of water. Calculate the dissolved carbon dioxide of this solution in parts per million. Your response must include both a correct numerical setup and the calculated result.

$$ppm = \frac{0.0023}{500.0023} \times 1,000,000$$

$$4.6 \ ppm$$

10. If 0.089 gram of KI is dissolved in 500. grams of H_2O, what is the concentration of the resulting solution in parts per million?

11. The dissolving of solid lithium bromide in water is represented by the balanced equation below.

$$LiBr(s) \xrightarrow{H_2O} Li^+(aq) + Br^-(aq)$$

Calculate the total mass of LiBr(s) required to make 500.0 grams of an aqueous solution of LiBr that has a concentration of 388 parts per million. Your response must include both a correct numerical setup and the calculated result.

1. 1   The mass of the solute (Na_3PO_4) is 0.02 g. The mass of solution is the mass of the solute + the mass of the solvent, or 1000.02 g.

    Substituting: parts per million $= \dfrac{0.02 \text{ g}}{1000.02 \text{ g}} \times 1000000$

    Solving: Concentration = 20 ppm

2. 4   The mass of the solute (NaCl) is 0.089 g. The mass of solution is the mass of the solute + the mass of the solvent, or 250.089 g.

    Substituting: parts per million $= \dfrac{0.089 \text{ g}}{250.089 \text{ g}} \times 1000000$

    Solving: Concentration = 356 ppm

3. Numerical setup: parts per million $= \dfrac{0.0070 \text{ g of } O_2}{1000. \text{ g of water} + 00.0070 \text{ g of } O_2} \times 1000000$

    or parts per million $= \dfrac{0.0070}{1000.0070} \times 1000000$

    Calculated result: Concentration = 7 ppm Significant figures need not to be shown.

    Explanation: The mass of the solute (oxygen) is 0.0070 g. The mass of solution is the mass of the solute + the mass of the solvent, or 1000.0070 g.

    Substituting: parts per million $= \dfrac{0.0070 \text{ g}}{1000.0070 \text{ g}} \times 1000000$

    Solving: Concentration = 7 ppm

4. Answer: 1.43 ppm   *or*   1.4 ppm

    Explanation: The mass of the solute (fluoride ions) is 0.000250 g. The mass of solution is the mass of the solute + the mass of solvent, or 175.000250 g.

    Substituting: parts per million $= \dfrac{0.000250 \text{ g}}{175.000250 \text{ g}} \times 1000000$

    Solving: Concentration = 1.43 ppm   *or* 1.4 ppm   Significant figures need not to be shown.

5. Answer: No

    Explanation: The mass of the solute (the minerals) is 0.04 g. The mass of solution (the water) is 750 g.

    Substituting: ppm $= \dfrac{0.04 \text{ g}}{750 \text{ g}} \times 1000000$ Solving: Concentration = 53 ppm

    This amount is under the hard water limit.

1. Molarity is defined as the

   (1) moles of solute per kilogram of solvent
   (2) moles of solute per liter of solution
   (3) mass of a solution
   (4) volume of a solvent   1 _____

2. A 3.0 M HCl(aq) solution contains a total of

   (1) 3.0 grams of HCl per liter of water
   (2) 3.0 grams of HCl per mole of solution
   (3) 3.0 moles of HCl per liter of solution
   (4) 3.0 moles of HCl per mole of water
                                     2 _____

3. What is the molarity of a solution that contains 0.50 mole of NaOH in 0.25 liter of solution?

   (1) 1.0 M   (3) 0.25 M
   (2) 2.0 M   (4) 0.50 M   3 _____

4. What is the molarity of 1.5 liters of an aqueous solution that contains 52 grams of lithium fluoride, LiF, (gram-formula mass = 26 grams/mole)?

   (1) 1.3 M   (3) 3.0 M
   (2) 2.0 M   (4) 0.75 M   4 _____

5. What is the total number of moles of NaCl(s) needed to make 3.0 liters of a 2.0 M NaCl solution?

   (1) 1.0 mol   (3) 6.0 mol
   (2) 0.70 mol   (4) 8.0 mol   5 _____

6. What is the total number of moles of solute in 4.0 liters of 4.0 M NaOH?

   (1) 1.0 mole   (3) 3.0 moles
   (2) 2.0 moles   (4) 16 moles   6 _____

Base your answers to question 7 on the information below.

A student is instructed to make 0.250 liter of a 0.200 M aqueous solution of $Ca(NO_3)_2$.

7. *a*) In the space below, show a correct numerical setup for calculating the total number of moles of $Ca(NO_3)_2$ needed to make 0.250 liter of the 0.200 M calcium nitrate solution.

   b) In order to prepare the described solution in the laboratory, two quantities must be measured accurately. One of these quantities is the volume of the solution. What other quantity must be measured to prepare this solution?

8. Which unit can be used to express solution concentration?

    (1) J/mol   (3) mol/L
    (2) L/mol   (4) mol/s

    8 _____

9. What is the molarity of a solution containing 20 grams of NaOH in 500 milliliters of solution?

    $\frac{20}{40g} \cdot 2 = 0.5\,mol,$

    $.5/.5 =$

    (1) 1 M   (3) 0.04 M
    (2) 2 M   (4) 0.5 M

    9 _____

10. What is the total number of moles of solute in 250 milliliters of a 1.0 M solution of NaCl?

    $\times/.250 = 1$

    (1) 1.0 mole   (3) 0.50 mole
    (2) 0.25 mole (4) 42 moles

    10 _____

11. What is the total number of grams of NaI(s) needed to make 1.0 liter of a 0.010 M solution?

    $.01 = \frac{\times}{1}$

    (1) 0.015   (3) 1.5
    (2) 0.15   (4) 15

    11 _____

12. A student wants to prepare a 1.0-liter solution of a specific molarity. The student determines that the mass of the solute needs to be 30. grams. What is the proper procedure to follow?

    (1) Add 30. g of solute to 1.0 L of solvent.
    (2) Add 30. g of solute to 970. mL of solvent to make 1.0 L of solution.
    (3) Add 1000. g of solvent to 30. g of solute.
    (4) Add enough solvent to 30. g of solute to make 1.0 L of solution.

    12 _____

13. Which preparation produces a 2.0 M solution of $C_6H_{12}O_6$? [molecular mass = 180.0]

    (1) 90.0 g of $C_6H_{12}O_6$ dissolved in 500.0 mL of solution
    (2) 90.0 g of $C_6H_{12}O_6$ dissolved in 1000. mL of solution
    (3) 180.0 g of $C_6H_{12}O_6$ dissolved in 500.0 mL of solution
    (4) 180.0 g of $C_6H_{12}O_6$ dissolved in 1000. mL of solution

    13 _____

14. In the space provided below, show a correct numerical setup for determining how many liters of a 1.2 M solution can be prepared with 0.50 mole of $C_6H_{12}O_6$.

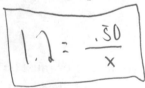

$$1.2 = \frac{.50}{x}$$

15. In the space provided below, show a correct numerical setup for calculating the total number of moles of ethylene glycol needed to prepare 2.50 liters of a 10.0 M solution.

$$10 = \frac{x}{2.5}$$

1. 2   Open to Table T and locate the Concentration row. Here it shows that $molarity = \dfrac{moles\ of\ solute}{liters\ of\ solution}$.

2. 3   By definition, the molarity of a solution is the number of moles of solute in 1 liter of solution. This is given in Table T – Concentration row, molarity section.

3. 2   The variables given are: moles of NaOH = 0.50 moles, liters of solution = 0.25 L.

   Equation: $molarity = \dfrac{moles\ of\ solute}{liters\ of\ solution}$

   Substituting: $molarity = \dfrac{0.5\ mole\ \ (NaOH)}{0.25\ liters\ of\ solution}$ Solving: molarity = 2.0 M

4. 1   Open to Table T and find the Mole Calculations equation (page 184) and in the Concentration section (page 197), locate the molarity equation. Using the first equation, number of moles = (52 g)/(26 g/mol) = 2.0 mol. Using the molarity equation = (2.0 mol)/(1.5 L) = 1.3 M.

5. 3   The variables given are: molarity = 2.0 M, liters of solution = 3.0 L

   Equation: From the molarity equation, moles of solute = (molarity)(liters of solution).

   Substituting: moles of NaCl $= (2.0\ \dfrac{mol}{L})(3.0\ L)$

   Solving: moles of NaCl = 6.0 mol

6. 4   The variables given are: molarity = 4.0 M, liters of solution = 4.0 L

   Equation: From the molarity equation, moles of solute = (molarity)(liters of solution).

   Substituting: moles of NaOH $= (4.0\ \dfrac{mol}{L})(4.0\ L)$

   Solving: moles of NaOH = 16 mol

7. a)   Numerical setup: moles of $Ca(NO_3)_2$ = (0.250 L)(0.200 M)   or $(0.250\ L)\dfrac{(0.200\ mol)}{(1\ L)}$

      or $0.2 = \dfrac{x}{0.25}$

   Equation: $molarity = \dfrac{moles\ of\ solute}{liters\ of\ solution}$

   Solving: $0.200\ \dfrac{mol}{L} = \dfrac{x}{0.250\ L}$, $x = (0.250\ L)\dfrac{(0.200\ mol)}{(1\ L)}$

   b)   Answer: mass of $Ca(NO_3)_2$ or   mass of solute   or   mass

   Explanation: To prepare a solution of given molarity, the number of moles of solute must be measured in addition to the volume of the solution. Therefore, the mass of $Ca(NO_3)_2$, the solute, must be measured to obtain the required number of moles.

$$\frac{P_1 V_1}{T_1} = \frac{P_2 V_2}{T_2}$$

P = pressure
V = volume
T = temperature (K)

Overview:

The volume of a given mass of gas depends upon the temperature and pressure of the sample of gas. Given an initial set of conditions, represented by the subscript 1, this law enables one to calculate the new volume of the gas when the temperature and/or pressure changes, represented by the subscript 2. Note that the temperature must be expressed on the Kelvin (K) temperature scale. If standard temperature or pressure are used (STP), refer to Table A for their values.

Example:

When a car air bag inflates, the nitrogen gas is at a pressure of 1.30 atmospheres, a temperature of 301 K, and has a volume of 40.0 liters. Calculate the volume of the nitrogen gas at STP. Your response must include both a correct numerical setup and the calculated volume.

Solution: From Table A, STP is 1 atm and 273 K.
The unknown is the final volume (V_2) of the nitrogen gas. The known quantities are:
P_1 = 1.30 atm, T_1 = 301 K, V_1 = 40.0 L, T_2 = 273 K, P_2 = 1 atm.

Equation: $V_2 = \dfrac{T_2 P_1 V_1}{T_1 P_2}$

Numerical setup: $V_2 = \dfrac{(273 \text{ K})(1.30 \text{ atm})(40.0 \text{ L})}{(301 \text{ K})(1.00 \text{ atm})}$

Calculated volume: V_2 = 47.2 L

Additional Information:

- If the temperature or pressure is unchanged, it can be left out of the equation.

- At constant temperature, the volume of a gas varies inversely with the pressure exerted on it.

- At constant pressure, the volume of a gas varies directly with the temperature.

- Ideal gas conditions are best achieved at high temperature and low pressure.

1. Standard pressure is equal to

   (1) 1 atm   (3) 273 atm

   (2) 1 kPa   (4) 273 kPa   1 _____

2. The volume of a gas is 4.00 liters at 293 K and constant pressure. For the volume of the gas to become 3.00 liters, the Kelvin temperature must be equal to

   (1) $\dfrac{3.00 \times 293}{4.00}$ (3) $\dfrac{3.00 \times 4.00}{293}$

   (2) $\dfrac{4.00 \times 293}{3.00}$ (4) $\dfrac{293}{3.00 \times 4.00}$ 2 _____

3. Which graph best represents the pressure-volume relationship for an ideal gas at constant temperature?

   (1)   (3)

   (2)   (4)   3 _____

4. As the temperature of a gas increases at constant pressure, the volume of the gas

   (1) decreases

   (2) increases

   (3) remains the same   4 _____

5. A sample of helium gas has a volume of 900. milliliters and a pressure of 2.50 atm at 298 K. What is the new pressure when the temperature is changed to 336 K and the volume is decreased to 450. milliliters?

   (1) 0.177 atm   (3) 5.64 atm

   (2) 4.43 atm   (4) 14.1 atm   5 _____

6. A rigid cylinder with a movable piston contains a 2.0-liter sample of neon gas at STP. What is the volume of this sample when its temperature is increased to 30.°C while its pressure is decreased to 90. kilopascals?

   (1) 2.5 L   (3) 1.6 L

   (2) 2.0 L   (4) 0.22 L   6 _____

7. A sample of oxygen gas in one container has a volume of 20.0 milliliters at 297 K and 101.3 kPa. The entire sample is transferred to another container where the temperature is 283 K and the pressure is 94.6 kPa. In the space below, show a correct numerical setup for calculating the new volume of this sample of oxygen gas.

Base your answer to question 8 on the information below.

Air bags are an important safety feature in modern automobiles. An air bag is inflated in milliseconds by the explosive decomposition of $NaN_3(s)$. The decomposition reaction produces $N_2(g)$, as well as $Na(s)$, according to the unbalanced equation below.

$$NaN_3(s) \rightarrow Na(s) + N_2(g)$$

8. When the air bag inflates, the nitrogen gas is at a pressure of 1.30 atmospheres, a temperature of 301 K, and has a volume of 40.0 liters. In the space below, calculate the volume of the nitrogen gas at STP. Your response must include both a correct numerical setup and the calculated volume.

Base your answer to question 9 using the information and diagrams below and your knowledge of chemistry.

Cylinder A contains 22.0 grams of $CO_2(g)$ and cylinder B contains $N_2(g)$. The volumes, pressures, and temperatures of the two gases are indicated under each cylinder.

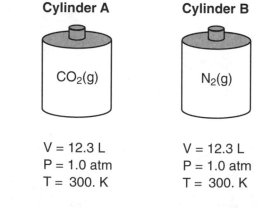

Cylinder A

$CO_2(g)$

V = 12.3 L
P = 1.0 atm
T = 300. K

Cylinder B

$N_2(g)$

V = 12.3 L
P = 1.0 atm
T = 300. K

9. The temperature of the $CO_2(g)$ is increased to 450. K and the volume of cylinder A remains constant. In the space below, show a correct numerical setup for calculating the new pressure of the $CO_2(g)$ in cylinder A.

Important Formulas and Equations

10. As the pressure of a given sample of a gas decreases at constant temperature, the volume of the gas

    (1) decreases
    (2) increases
    (3) remains the same   10 ____

11. A gas occupies a volume of 40.0 milliliters at 20°C. If the volume is increased to 80.0 milliliters at constant pressure, the resulting temperature will be equal to

    (1) $20°C \times \dfrac{80.0\,mL}{40.0\,mL}$ (3) $293\,K \times \dfrac{80.0\,mL}{40.0\,mL}$

    (2) $20°C \times \dfrac{40.0\,mL}{80.0\,mL}$ (4) $293\,K \times \dfrac{40.0\,mL}{80.0\,mL}$

    11 ____

12. A gas occupies a volume of 444 mL at 273 K and 79.0 kPa. What is the final kelvin temperature when the volume of the gas is changed to 1,880 mL and the pressure is changed to 38.7 kPa?

    (1) 31.5 K   (3) 566 K
    (2) 292 K   (4) 2,360 K   12 ____

13. A sample of gas occupies a volume of 50.0 milliliters in a cylinder with a movable piston. The pressure of the sample is 0.90 atmosphere and the temperature is 298 K. What is the volume of the sample at STP?

    (1) 41 mL   (3) 51 mL
    (2) 49 mL   (4) 55 mL   13 ____

14. A lightbulb contains argon gas at a temperature of 295 K and at a pressure of 75 kilopascals. The lightbulb is switched on, and after 30 minutes its temperature is 418 K.

    In the space below, show a correct numerical setup for calculating the pressure of the gas inside the lightbulb at 418 K. Assume the volume of the lightbulb remains constant.

15. A sample of helium gas is in a closed system with a movable piston. The volume of the gas sample is changed when both the temperature and the pressure of the sample are increased. The table shows the initial temperature, pressure, and volume of the gas sample, as well as the final temperature and pressure of the sample.

    Helium Gas in a Closed System

Condition	Temperature (K)	Pressure (atm)	Volume (mL)
initial	200.	2.0	500.
final	300.	7.0	?

    In the space below, show a correct numerical setup for calculating the final volume of the helium gas sample.

Base your answers to question 16 using the information below and your knowledge of chemistry.

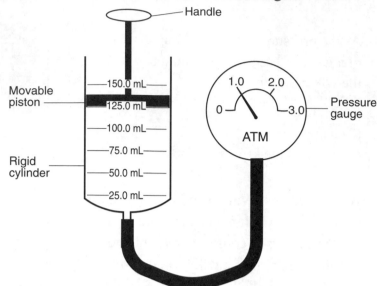

A rigid cylinder is fitted with a movable piston. The cylinder contains a sample of helium gas, He(g), which has an initial volume of 125.0 milliliters and an initial pressure of 1.0 atmosphere, as shown. The temperature of the helium gas sample is 20.0°C.

16. *a)* Express the initial volume of the helium gas sample in liters.

_____

b) The piston is pushed further into the cylinder. In the space below, show a correct numerical setup for calculating the volume of the helium gas that is anticipated when the reading on the pressure gauge is 1.5 atmospheres. The temperature of the helium gas remains constant.

Base your answers to question 17 using the information below and your knowledge of chemistry.

A weather balloon has a volume of 52.5 liters at a temperature of 295 K. The balloon is released and rises to an altitude where the temperature is 252 K.

17. *a)* How does this temperature change affect the gas particle motion?

_____

b) The original pressure at 295 K was 100.8 kPa and the pressure at the higher altitude at 252 K is 45.6 kPa. Assume the balloon does not burst. In the space below, show a correct numerical setup for calculating the volume of the balloon at the higher altitude.

1. 1   Table A (page 1) gives the standard pressure as 101.3 kPa or 1 atm.

2. 1   In this problem the pressure is constant. Therefore $P_1 = P_2$ and may be eliminated from the Combined Gas Law equation. Therefore $\dfrac{V_1}{T_1} = \dfrac{V_2}{T_2}$

   The known values are: $V_1 = 4.00$ L, $T_1 = 293$ K, $V_2 = 3.00$ L and T_2 is the unknown.

   Substituting : $\dfrac{4.00\ \text{L}}{293\ \text{K}} = \dfrac{3.00\ \text{L}}{T_2}$

   Solving: $T_2 = \dfrac{(3.00\ \text{L})(293\ \text{K})}{4.00\ \text{L}}$

3. 4   The volume of a gas varies inversely with the pressure at constant temperature. Graph 4 shows this relationship.

4. 2   At constant pressure, the volume of a given mass of a gas varies directly with temperature. Thus, if the pressure remains the same, an increase in temperature of a gas will cause an increase in the volume of the gas.

5. 3   The variables given are: $V_1 = 900.$ mL, $P_1 = 2.50$ atm, $T_1 = 298$ K, $T_2 = 336$ K, $V_2 = 450.$ mL, P_2 is the unknown.

   Substituting into the Combined Gas Law equation: $\dfrac{(2.50\ \text{atm})(900.\ \text{mL})}{298\ \text{K}} = \dfrac{(P_2)(450.\ \text{mL})}{336\ \text{K}}$

   Solving: $P_2 = \dfrac{(2.50\ \text{atm})(900.\ \text{mL})(336\ \text{K})}{(298\ \text{K})(450.\ \text{mL})} = 5.64$ atm

6. 1   The variables given are: $V_1 = 2.0$ L, $T_2 = 303$ K (30.°C), $P_2 = 90.$ kPa and at STP, $P_1 = 101.3$ kPa and $T_1 = 273$ K (0°C).

   Substituting and solving: $V_2 = \dfrac{(101.3\ \text{kPa})\ (2.0\ \text{L})(303\ \text{K})}{(273\ \text{K})(90.\ \text{kPa})}$ $V_2 = 2.5$ L

7. Numerical setup is: $V_2 = \dfrac{(101.3\ \text{kPa})(20.0\ \text{mL})(283\ \text{K})}{(297\ \text{K})\ (94.6\ \text{kPa})}$ *or* $\dfrac{(101.3)(20.0)}{297} = \dfrac{(94.6)V_2}{283}$

   Explanation: The new volume would be V_2. Substitute the given variables into the Combined Gas Law equation and solve for V_2. The correct numerical setups are shown above.

8. Numerical setup: $V_2 = \dfrac{(273\ \text{K})(1.30\ \text{atm})(40.0\ \text{L})}{(301\ \text{K})(1.00\ \text{atm})}$

or

$$V_2 = \dfrac{(273)(1.30)(40.0)}{(301)(1.00)}$$

or

$$V_2 = \dfrac{(1.30\ \text{atm})(40.0\ \text{L})}{310\ \text{K}} = \dfrac{(1\ \text{atm})\,V_2}{273\ \text{K}}$$

Answer: 47.2 L   Significant figures do not need to be shown.

Explanation: STP is 1 atm, which is P_2, and 273 K, which is T_2 (see Table A). The other variables are: $V_1 = 40.0$ L, $T_1 = 301$ K.

Solving for V_2 in the Combined Gas Law equation gives:

$V_2 = \dfrac{(T_2)(P_1)(V_1)}{(T_1)(P_2)}$ Substituting into this equation gives the numerical setup shown above.

Solving: $V_2 = \dfrac{14{,}196\ \text{L}}{301} = 47.16$ L *or* 47.2 L

9. Numerical setup: $\dfrac{(1.0\ \text{atm})(12.3\ \text{L})}{300\ \text{K}} = \dfrac{(P_2)(12.3\ \text{L})}{450\ \text{K}}$ *or* $\dfrac{1}{300} = \dfrac{P_2}{450}$

or $\dfrac{1.0\ \text{atm}}{300\ \text{K}} = \dfrac{P_2}{450\ \text{K}}$ *or* $P_2 = \dfrac{(1.0\ \text{atm})(450\ \text{K})}{300\ \text{K}}$

Explanation: The volume remains constant, therefore may be eliminated from the Combined Gas Law. Substituting the known variables directly into the Combined Gas Law equation gives the setup shown above.

Equation: $\dfrac{P_1}{T_1} = \dfrac{P_2}{T_2}$ P_2 is the unknown.

Solving: $P_2 = \dfrac{(P_1)(T_2)}{(T_1)}$

Substituting: $P_2 = \dfrac{(1.0\ \text{atm})(450\ \text{K})}{300\ \text{K}}$

Important Formulas and Equations

$$M_A V_A = M_B V_B$$

M_A = molarity of H⁺	M_B = molarity of OH⁻
V_A = volume of acid	V_B = volume of base

Overview:

Titration is a process used to determine the concentration (molarity) of an acid or base by slowly combining it with a base or acid, respectively, of known concentration, called a standard.

Example:

What volume of 0.250 M HCl(aq) must completely react to neutralize 50.0 milliliters of 0.100 M NaOH(aq)?

Equation: $M_A V_A = M_B V_B$

Given values: M_A = 0.250 M HCl(aq), V_A = unknown, M_B = 0.100 M NaOH(aq) and V_B = 50.0 mL.

Substitution: $(0.250 \text{ M})(V_A) = (0.100 \text{ M})(50.0 \text{ mL})$

Answer: V_A of HCl(aq) = 20.0 mL

Additional Information:

- An indicator can be used to determine the end point in an acid-base titration. At the end point, the resulting solution is not necessarily neutral (pH = 7). It depends on the strength of the acid and base.

- The net ionic equation for a neutralization reaction is $H^+(aq) + OH^-(aq) \rightarrow H_2O(\ell)$.

- If the number of H⁺ in the acid formula and the number of OH⁻ in the base formula are the same, the calculated concentration of the acid or base is the correct concentration.

- When calculating the concentration of an acid, if the acid has 2 H⁺ and the base has 1 OH⁻, divide the calculated concentration by 2 to obtain the correct concentration of the acid, See Set 1, question 4.

- When calculating the concentration of a base, if the base has 2 OH⁻ and the acid has 1 H⁺, divide the calculated concentration by 2 to obtain the correct concentration of the base.

1. The diagram represents a section of a buret containing acid used in an acid-base titration. What is the total volume of acid that was used?

   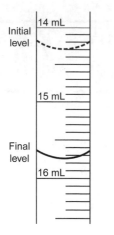

   (1) 1.10 mL
   (2) 1.30 mL
   (3) 1.40 mL
   (4) 1.45 mL

   1 _____

2. If 5.0 milliliters of a 0.20 M HCl solution is required to neutralize exactly 10. milliliters of NaOH, what is the concentration of the base?

   (1) 0.10 M   (3) 0.30 M
   (2) 0.20 M   (4) 0.40 M   2 _____

3. What volume of 0.500 M HNO_3(aq) must completely react to neutralize 100.0 milliliters of 0.100 M KOH(aq)?

   (1) 10.0 mL   (3) 50.0 mL
   (2) 20.0 mL   (4) 500. mL   3 _____

4. Information related to a titration experiment is given in the balanced equation and table below.

   Titration Experiment Results

volume of H_2SO_4(aq) used	12.0 mL
concentration of H_2SO_4(aq)	?
volume of KOH(aq) used	36.0 mL
concentration of KOH(aq)	0.16 M

   $$H_2SO_4(aq) + 2KOH(aq) \rightarrow K_2SO_4(aq) + 2H_2O(\ell)$$

   Based on the equation and the titration results, what is the concentration of the H_2SO_4(aq)?

   (1) 0.12 M   (3) 0.24 M
   (2) 0.16 M   (4) 0.96 M   4 _____

5. Samples of acid rain are brought to a laboratory for analysis. Several titrations are performed and it is determined that a 20.0-milliliter sample of acid rain is neutralized with 6.50 milliliters of 0.010 M NaOH. What is the molarity of the H^+ ions in the acid rain?

Base your answers to question 6 using the information below and your knowledge of chemistry.

   A student titrates 60.0 mL of HNO_3(aq) with 0.30 M NaOH(aq). Phenolphthalein is used as the indicator. After adding 42.2 mL of NaOH(aq), a color change remains for 25 seconds, and the student stops the titration.

6. *a)* What color change does phenolphthalein undergo during this titration? _____

   b) Show a correct numerical setup for calculating the molarity of the HNO_3(aq).

Important Formulas and Equations

7. A student recorded the following buret readings during a titration of a base with an acid:

	Standard 0.100 M HCl	Unknown KOH
Initial reading	9.08 mL	0.55 mL
Final reading	19.09 mL	5.56 mL

a) Calculate the molarity of the KOH. Show all work.

b) Record your answer to the correct number of significant figures. _____

Base your answers to question 8 using the information below and your knowledge of chemistry.

In a titration, 3.00 M NaOH(aq) was added to an Erlenmeyer flask containing 25.00 milliliters of HCl(aq) and three drops of phenolphthalein until one drop of the NaOH(aq) turned the solution a light-pink color. The following data were collected by a student performing this titration:

Initial NaOH(aq) buret reading: 14.45 milliliters
Final NaOH(aq) buret reading: 32.66 milliliters

8. a) What is the total volume of NaOH(aq) that was used in this titration? _____ mL

b) Show a correct numerical setup for calculating the molarity of the HCl(aq).

c) Based on the data given, what is the correct number of significant figures that should be shown in the molarity of the HCl(aq)? _____

Important Formulas and Equations

$$M_a V_a = M_B V_B$$

Set 2 — Titration

9. The data collected from a laboratory titration are used to calculate the

(1) rate of a chemical reaction
(2) heat of a chemical reaction
(3) concentration of a solution
(4) boiling point of a solution   9 _____

10. If 5.0 milliliters of a 0.20 M HCl solution is required to neutralize exactly 10. milliliters of NaOH, what is the concentration of the base?

Acid

$(.20)(5) = (10)(x)$
$1 = 10x$
$\frac{1}{10} = \frac{10x}{10}$
$1 = .10x$

(1) 0.10 M   (3) 0.30 M
(2) 0.20 M   (4) 0.40 M   10 __1__

11. How many milliliters of 0.20 M KOH are needed to completely neutralize 90.0 milliliters of 0.10 M HCl?

Acid

(1) 25 mL   (3) 90. mL
(2) 45 mL   (4) 180 mL   11 __2__

$(.20)(x) = (90)(.10)$
$\frac{20x}{20} = \frac{9}{20}$
Base   *Acid*

12. When 50. milliliters of an HNO_3 solution is exactly neutralized by 150 milliliters of a 0.50 M solution of KOH, what is the concentration of HNO_3? *Base*

(1) 1.0 M   (3) 3.0 M
(2) 1.5 M   (4) 0.5 M   x=1   12 __2__

$(x)(50) = (.50)(150)$

Base your answers to question 13 using the information below and your knowledge of chemistry.

Using burets, a student titrated a sodium hydroxide solution of unknown concentration with 0.10 M HCL.

13. a) Determine both the total volume of HCl(aq) and the total volume of NaOH(aq) used in the titration.

HCl(aq) = __9.5__ mL

Na(OH)(aq) = __3.8__ mL

Titration Data

Solution	HCl(aq)	NaOH(aq)
Initial Buret Reading (mL)	15.50	5.00
Final Buret Reading (mL)	25.00	8.80

Initial + Final =
Subtract

```
 25.4      8.80
-15.5      - 5
 9.5       3.8
```

b) Show a correct numerical setup for calculating the molarity of the sodium hydroxide solution.

Acid   *Base*
$(.10)(9.5) = (x)(3.8)$

c) Solve for the molarity of the sodium hydroxide solution.

$(.10)(9.5) = (x)(3.8)$

$.95 = 3.8x$
$\frac{.95}{3.8} = \frac{3.8x}{3.8}$

$\frac{.95}{3.8} = \frac{x \cdot 3.8}{3.8}$ x = .25

Important Formulas and Equations

Base your answers to question 14 using the information below and your knowledge of chemistry.

A 20.0-milliliter sample of HCl(aq) is completely neutralized by 32.0 milliliters of 0.50 M KOH(aq). _Acid_

14. *a)* Identify the negative ion produced when the KOH is dissolved in distilled water. ___OH^-___

b) In the space below, Calculate the molarity of the HCl(aq). Your response must include both a numerical setup and the calculated result.

c) Complete the equation representing this titration reaction by writing the formulas of the products.

HCl(aq) + KOH(aq) → ___KCl___ + ___H_2O___

Acid   _Base_

Base your answers to question 15 on the information below.

In preparing to titrate an acid with a base, a student puts on goggles and an apron. The student uses burets to dispense and measure the acid and the base in the titration. In each of two trials, a 0.500 M NaOH(aq) solution is added to a flask containing a volume of HCl(aq) solution of unknown concentration. Phenolphthalein is the indicator used in the titration. The calculated volumes used for the two trials are recorded in the table below.

Volumes of Base and Acid Used in Titration Trials

Solution (aq)	Molarity (M)	Trial 1 Volume Used (mL)	Trial 2 Volume Used (mL)
NaOH	0.500	17.03	16.87
HCl	?	10.22	10.12

Don't Subtract

15. *a)* Write a chemical name for the acid used in the titration.

___Hydrochloric acid___

b) Using the volumes from trial 1, determine the molarity of the HCl(aq) solution.

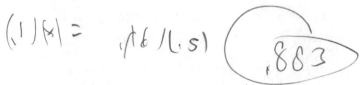

c) Based on the information given in the table, how many significant figures should be shown in the calculated molarity of the HCl(aq) solution used in Trial 2? ___5___

1. **4** Subtracting the final level from the initial level will give the total volume of acid used in the titration.

   Readings: final level = 15.75 mL, initial level = 14.30 mL. The difference of these readings is 1.45 mL.

2. **1** In an acid-base titration procedure, a solution of known concentration (the standard) is used to determine the unknown concentration of an acid or base by reaching neutralization.

   The given acid is HCl and the given base is NaOH.

   Equation: $M_A V_A = M_B V_B$

   Given values: $M_A = 0.20$ M, $V_A = 5.0$ mL, $V_B = 10.$ mL, M_B is the unknown

   Substituting: $(0.20$ M$)(5.0$ mL$) = (M_B)(10.$ mL$)$

   Solving: $M_B = \dfrac{(5.0 \text{ mL})(0.20 \text{ M})}{10. \text{ mL}} = 0.10$ M

3. **2** The given acid is HNO_3 and the given base is KOH.

   Given values: $M_A = 0.500$ M, $M_B = 0.100$ M, $V_B = 100.0$ mL, V_A is the unknown

   Equation: $M_A V_A = M_B V_B$

   Substituting: $(0.500$ M$)(V_A) = (0.100$ M$)(100.0$ mL$)$

   Solving: $V_A = \dfrac{(0.100 \text{ M})(100.0 \text{ mL})}{0.500 \text{ M}} = 20.0$ mL

4. **3** The given acid is H_2SO_4 and the given base is KOH.

   Given values: $V_A = 12.0$ mL, $M_B = 0.16$ M, $V_B = 36.0$ mL, M_A is the unknown

   Equation: $M_A V_A = M_B V_B$

   Substituting: $(M_A)(12.0$ mL$) = (0.16$ M$)(36.0$ mL$)$

   Solving: $M_A = 0.48$ M   Dividing by 2 gives: 0.48 M/2 = 0.24 M

   Note: The acid is a diprotic acid (two moles of H^+ per mole of acid) and the base is a monohydroxy base (one mole of OH^- per mole of base). Therefore, it will require a solution of acid only one-half the calculated concentration for this titration:
   $$\frac{0.48}{2} = 0.24 \text{ M}$$

   In a titration, the neutralization equation is: $H^+ + OH^- \rightarrow HOH$ (1:1 ratio for H^+ to OH^-).

Important Formulas and Equations

5. Answer: 0.0033 M   *or* 3.3×10^{-3} M   *or* 3.25×10^{-3} M

   Explanation: The acid-base titration method is used to determine the unknown concentration (molar concentration) of an acid or base.

   Given values: V_A = 20.0 mL, M_B = 0.010 M, V_B = 6.50 mL, M_A = is the unknown for molarity of H^+ ions

   Equation: $M_A V_A = M_B V_B$

   Substituting: (M_A)(20.0 mL) = (0.010 M)(6.50 mL)

   Solving: M_A of H^+ ions = 0.0033 M

6. *a)* Answer: Colorless to pink

   Explanation: Phenolphthalein is colorless in the HNO_3(aq) solution. When NaOH(aq), a base (see Table L), is added to the HNO_3(aq), an acid (see Table K), the pH of the solution increases. When the pH of the solution becomes 8, the phenolphthalein will change to a pink color (see Table M).

   b) Answer: $M_A = \dfrac{(0.30\ M)(42.2\ mL)}{60.0\ mL}$ *or* $(\times)(60) = (.3)(42.2)$

   Explanation: Open to Table K and L. Here it identifies HNO_3 as nitric acid and NaOH as sodium hydroxide, a base.

   Given values: V_A = 60.0 mL of HNO_3, M_B = 0.30 M, V_B = 42.2 mL, M_A is the unknown molarity of HNO_3(aq)

   Equation: $M_A V_A = M_B V_B$

   Substituting: (M_A)(60.0 mL) = (0.30 M)(42.2 mL)

   Numerical setup: $M_A = \dfrac{(0.30\ M)(42.2\ mL)}{60.0\ mL}$

7. *a)* Answer: $M_B = \dfrac{(0.100\ M)(10.01\ mL)}{5.01\ mL}$ *or* $(0.100\ M)(10.01\ mL) = (M_B)(5.01\ mL)$

   Explanation: As shown in the chart, the acid is HCl and the base is KOH.

   Given values: V_A = 10.01 mL (19.09 mL – 9.08 mL), V_B = 5.01 mL (5.56 mL – 0.55 mL), M_A = 0.100 M, M_B = molarity of KOH is the unknown

   Equation: $M_A V_A = M_B V_B$

   Substituting: (0.100 M)(10.01 mL) = (M_B)(5.01 mL)

   Solving: $M_B = \dfrac{(0.100\ M)(10.01\ mL)}{5.01\ mL}$

7. *b)* Answer: $M_B = 0.200$ M (three significant figures)

Explanation: The answer must be expressed to the same number of significant figures as that in the least accurate measurement (that with the least number of significant figures). In this case, the number is 3, found in both 0.100 M and 5.01 mL. The number 10.01 mL contains 4 significant figures.

8. *a)* Answer: 18.21 mL

Explanation: Subtracting the final buret reading from the initial buret reading gives the volume of base used, 32.66 mL – 14.45 mL = 18.21 mL.

b) Numerical setup: $M_A = \dfrac{(3.00 \text{ M})(18.21 \text{ mL})}{25.00 \text{ mL}}$ or $M_A(25) = (3)(18.21)$

Explanation: As shown in Table K and L, NaOH(aq) is a base and HCl(aq) is an acid. The variables given are: $M_B = 3.00$ M, $V_A = 25.00$ mL, $V_B = 18.21$ mL, $M_A =$ is the unknown for HCl(aq).

Substituting into the titration equation gives you the setup shown above.

c) Answer: 3

Explanation: Refer to the rule given in the explanation for question 7*b*. In this case, 3.00 M has 3 significant figures and 18.21 mL and 25.00 mL both have 4 significant figures. The least number of significant figures in these measurements is 3. Therefore, the answer must be rounded to 3 significant figures.

$$q = mC\Delta T$$
$$q = mH_f$$
$$q = mH_v$$

q = heat

m = mass

C = specific heat capacity

ΔT = change in temperature

H_f = heat of fusion

H_v = heat of vaporization

Overview:

When heat is added to or removed from a substance, two things can occur.

1. If the substance is a solid or liquid not at its melting point or boiling point, respectively, its temperature will change. In this case, use the equation $q = mC\Delta T$, where ΔT is in Kelvin.

2. If the substance is at its melting point or boiling point, a change of phase will occur. For melting or freezing, use the equation $q = mH_f$. For vaporization or condensation, use the equation $q = mHv$. Since there is no temperature change during a change of phase, there is no ΔT term in these equations. For water, the values of C, H_f and Hv are given on Table B.

Examples:

(1) How much heat is needed to bring 25 grams of water from an initial temperature of 293 K to 338 K?

Solution: In this problem, water is increasing in temperature and not going through any phase change. The equation to use is $q = mC\Delta T$, where $m = 25$ grams, $C = 4.18$ J/g•K (see Table B), and the $\Delta T = 45$ K (338 K – 293 K)

Substitution: $q = (25\ g)(4.18\ J/g•K)(45\ K)$

Answer: $q = 4{,}702.5$ J

(2) How much heat is needed to completely change 10 grams of ice to water at the melting point temperature?

Solution: This is a phase change of melting, so the heat of fusion equation is used.
$q = mH_f$, where $m = 10$ grams and H_f is 334 J/g (see Table B)

Substitution: $q = (10\ g)(334\ J/g)$

Answer: $q = 3{,}334$ J

(3) How much heat is needed to completely boil away 15 grams of water that is at its boiling point temperature?

Solution: This is a phase change of vaporization (boiling), so the heat of vaporization equation is used.
$q = mHv$, where $m = 15$ grams and Hv is 2,260 J/g (see Table B)

Substitution: $q = (15\ g)(2{,}260\ J/g)$

Answer: $q = 33{,}900$ J

Heating Curve:

A graph of temperature vs. time during which a substance heats up is known as a heating curve. A heating curve is helpful in illustrating what occurs as a substance changes from a solid to a liquid to a gas. A typical heating curve is shown to the right.

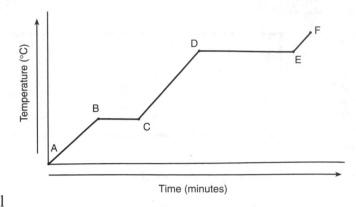

Notice that there are three sloped intervals and two horizontal intervals (the plateaus) on this temperature-time graph. The three-sloped interval indicates the added heat is causing an increase in temperature of the substance. The two plateaus represent heat causing a phase change of the substance. During a phase change, since there is no change of temperature, the average kinetic energy of the molecules remains the same, but the added heat increases the internal potential energy of the substance.

Explanation of the sections:

Section A-B: The solid substance is absorbing heat and is increasing in temperature. During this time, the kinetic energy of the molecules is increasing.

Section B-C: The phase change of melting is occurring. Since the temperature is constant and heat is still being supplied, the kinetic energy remains the same, but the potential energy is increasing.

Section C-D: This section represents the liquid substance heating up. Once again, since temperature is increasing, the kinetic energy of the molecules is increasing.

Section D-E: The phase change of boiling is occurring. Here the kinetic energy remains the same, but the potential energy is increasing.

Section E-F: The substance is now in the gaseous phase and its temperature is increasing. Again, the kinetic energy of the molecules in increasing.

From the length of the horizontal lines (the plateaus) on the graph, it shows that the heat of vaporization is greater than the heat of fusion.

In a *cooling curve*, the graph would show a substance decreasing in temperature with different intervals representing a decrease in temperature and horizontal intervals (plateaus) representing the phase changes of condensation and freezing of the substance. During these phase changes, the molecular kinetic energy remains the same, but the molecular potential energy is decreasing.

Remember, all phase transitions occur at a constant temperature, producing a plateau on a heating or cooling curve.

Important Formulas and Equations

Cooling Curve

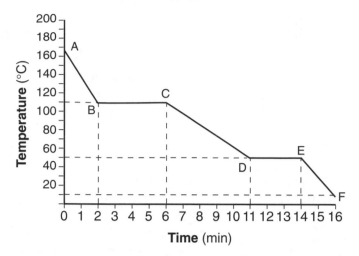

In this cooling curve, the initial temperature is above the substance's boiling point and its final temperature is below its freezing point. During the 16-minute interval, heat is being released, decreasing the average molecular kinetic energy in intervals AB, CD and EF and decreasing the molecular potential energy in intervals BC and DE.

Additional Information:

- The joule (J) is the unit for the quantity of heat in the Metric System, see Table D.

- Whenever two bodies with different temperatures get close enough to each other, heat flows from the body with the higher temperature to the body with the lower temperature.

- Sublimation is the change from the solid directly to a gas without going through the liquid phase. Dry ice, $CO_2(s)$, undergoes this process.

- Melting and boiling are endothermic reactions since they absorb heat.

- Freezing and condensation are exothermic reactions since they release heat.

- The specific heat capacity for water is the highest of all common substances.

- The celsius degree and the kelvin are the same size. Therefore, a given temperature change on the celsius scale and the kelvin scale will be the same. The unit for specific heat capacity may then be expressed as J/g•°C.

1. The heat absorbed when ice melts can be measured in a unit called a

   (1) torr   (3) mole
   (2) degree   (4) joule   1 _____

2. The heat energy required to change a unit mass of a solid into a liquid at constant temperature is called

   (1) heat of vaporization
   (2) heat of formation
   (3) heat of solution
   (4) heat of fusion   2 _____

3. What amount of heat is required to completely melt a 29.95-gram sample of $H_2O(s)$ at 0°C?

   (1) 334 J   (3) 1.00×10^3 J
   (2) 2,260 J   (4) 1.00×10^4 J   3 _____

4. How much energy is required to vaporize 10.00 grams of water at its boiling point?

   (1) 100 J   (3) 2,260 J
   (2) 260 J   (4) 22,600 J   4 _____

5. Two samples of gold that have different temperatures are placed in contact with one another. Heat will flow spontaneously from a sample of gold at 60°C to a sample of gold that has a temperature of

   (1) 50°C   (3) 70°C
   (2) 60°C   (4) 80°C   5 _____

6. The graph below represents the uniform heating of a substance, starting with the substance as a solid below its melting point.

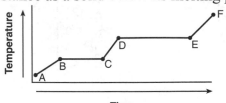

   Which line segment represents an increase in potential energy and no change in average kinetic energy?

   (1) $\overline{AB}$ (3) $\overline{CD}$
   (2) $\overline{BC}$ (4) $\overline{EF}$ 6 _____

7. The average kinetic energy of water molecules is greatest in which of these samples?

   (1) 10 g of water at 35°C
   (2) 10 g of water at 55°C
   (3) 100 g of water at 25°C
   (4) 100 g of water at 45°C   7 _____

8. The average kinetic energy of water molecules increases when

   (1) $H_2O(s)$ changes to $H_2O(\ell)$ at 0°C
   (2) $H_2O(\ell)$ changes to $H_2O(s)$ at 0°C
   (3) $H_2O(\ell)$ at 10°C changes to $H_2O(\ell)$ at 20°C
   (4) $H_2O(\ell)$ at 20°C changes to $H_2O(\ell)$ at 10°C

   8 _____

9. Which phase change is exothermic?

   (1) solid to liquid
   (2) solid to gas
   (3) liquid to solid
   (4) liquid to gas   9 _____

Important Formulas and Equations

10. *a*) In the space below, calculate the heat released when 25.0 grams of water freezes at 0°C. Show all work.

b) Record your answer with an appropriate unit. _____

Base your answers to question 11 using the accompanying heating curve, which represents a substance starting as a solid below its melting point and being heated at a constant rate over a period of time.

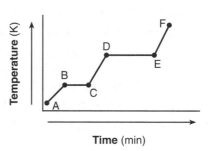

11. *a*) What is happening to the average kinetic energy of the particles during segment $\overline{BC}$?

b) How does this heating curve illustrate that the heat of vaporization is greater than the heat of fusion?

_____

_____

Base your answers to question 12 using the information below and your knowledge of chemistry.

A 5.00-gram sample of liquid ammonia is originally at 210. K. The diagram of the partial heating curve represents the vaporization of the sample of ammonia at standard pressure due to the addition of heat. The heat is not added at a constant rate. Some physical constants for ammonia are shown in the data table.

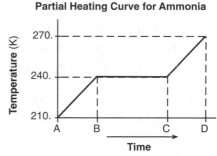

Partial Heating Curve for Ammonia

Some Physical Constants for Ammonia

specific heat capacity of $NH_3(\ell)$	4.71 J/g•K
heat of fusion	332 J/g
heat of vaporization	1370 J/g

12. *a*) Calculate the total heat absorbed by the 5.00-gram sample of ammonia during time interval *AB*. Your response must include both a correct numerical setup and the calculated result.

b) Determine the total amount of heat required to vaporize this 5.00-gram sample of ammonia at its boiling point.

13. As a solid substance absorbs heat at its melting point, the melting point will

    (1) decrease
    (2) increase
    (3) remain the same   13 ____

14. The heat energy required to change a unit mass of a solid into a liquid at constant temperature is called

    (1) heat of vaporization
    (2) heat of formation
    (3) heat of solution
    (4) heat of fusion   14 ____

15. An iron bar at 325 K is placed in a sample of water. The iron bar gains energy from the water if the temperature of the water is

    (1) 65 K   (3) 65°C
    (2) 45 K   (4) 45°C   15 ____

16. How much heat energy must be absorbed to completely melt 35.0 grams of $H_2O(s)$ at 0°C?

    (1) 9.54 J   (3) 11,700 J
    (2) 146 J   (4) 79,100 J   16 ____

17. What is the total number of joules released when a 5.00-gram sample of water changes from liquid to solid at 0°C?

    (1) 334 J   (3) 2,260 J
    (2) 1,670 J   (4) 11,300 J   17 ____

18. Which phase change is accompanied by the release of heat?

    (1) $H_2O(s) \rightarrow H_2O(g)$
    (2) $H_2O(s) \rightarrow H_2O(\ell)$
    (3) $H_2O(\ell) \rightarrow H_2O(g)$
    (4) $H_2O(\ell) \rightarrow H_2O(s)$ 18 ____

19. The temperature of a sample of water changes from 10°C to 20°C when the water absorbs 418 joules of heat. What is the mass of the sample?

    (1) 1 g   (3) 100 g
    (2) 10 g   (4) 1,000 g   19 ____

20. Which sample of Fe contains particles having the highest average kinetic energy?

    (1) 5 g at 10°C   (3) 5 g at 400 K
    (2) 10 g at 25°C   (4) 10 g at 300 K   20 ____

21. A 36-gram sample of water has an initial temperature of 22°C. After the sample absorbs 1,200 joules of heat energy, the final temperature of the sample is

    (1) 8.0°C   (3) 30.°C
    (2) 14°C   (4) 55°C   21 ____

22. What is the minimum amount of heat required to completely melt 20.0 grams of ice at its melting point?

    (1) 20.0 J   (3) 6,680 J
    (2) 83.6 J   (4) 45,200 J   22 ____

23. A person with a body temperature of 37°C holds an ice cube with a temperature of 0°C in a room where the air temperature is 20.°C. The direction of heat flow is

    (1) from the person to the ice, only
    (2) from the person to the ice and air, and from the air to the ice
    (3) from the ice to the person, only
    (4) from the ice to the person and air, and from the air to the person   23____

Important Formulas and Equations

24. What occurs when a 35-gram aluminum cube at 100.°C is placed in 90. grams of water at 25°C in an insulated cup?

(1) Heat is transferred from the aluminum to the water, and the temperature of the water decreases.
(2) Heat is transferred from the aluminum to the water, and the temperature of the water increases.
(3) Heat is transferred from the water to the aluminum, and the temperature of the water decreases.
(4) Heat is transferred from the water to the aluminum, and the temperature of the water increases.

24_____

Base your answers to question 25 on the information below.

Heat is added to a sample of liquid water, starting at 80.°C, until the entire sample is a gas at 120.°C. This process, occurring at standard pressure, is represented by the balanced equation below.

$$H_2O(\ell) + heat \rightarrow H_2O(g)$$

25. a) In the box, using the key, draw a particle diagram to represent at least four molecules of the product of this physical change at 120.°C.

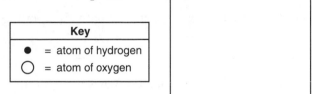

b) On the diagram, complete the heating curve for this physical change.

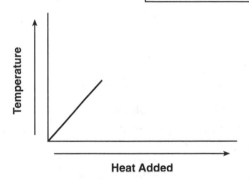

Base your answers to question 26 on the information below.

Heat is added to a 200. gram sample of $H_2O(s)$ to melt the sample at 273 K. Then the resulting $H_2O(\ell)$ is heated to a final temperature of 338 K.

26. a) Determine the total amount of heat required to completely melt the sample. _____

b) Show a numerical setup for calculating the total amount of heat required to raise the temperature of the $H_2O(\ell)$ from 273 K to its final temperature.

c) Compare the amount of heat required to vaporize a 200. gram sample of $H_2O(\ell)$ at its boiling point to the amount of heat required to melt a 200. gram sample of $H_2O(s)$ at its melting point.

_____

_____

27. The heat of fusion for a substance is 122 joules per gram. How many joules of heat are needed to melt 7.50 grams of this substance at its melting point? _____ J

28. A sample of water is heated from a liquid at 40°C to a gas at 110°C. The graph of the heating curve is shown.

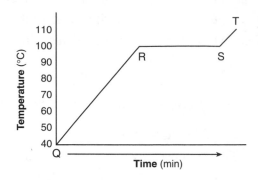

   a) On the accompanying heating curve diagram, label each of the following regions:

           Liquid, only

           Gas, only

           Phase change

   b) For section QR of the graph, state what is happening to the water molecules as heat is added.

_____

   c) For section RS of the graph, state what is happening to the water molecules as heat is added.

_____

Base your answers to question 29 on the information below.

    The accompanying graph shows a compound being cooled at a constant rate starting in the liquid phase at 75°C and ending at 15°C.

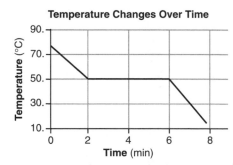

29. *a)* What is the freezing point of the compound, in degrees Celsius? _____ °C

   b) A different experiment was conducted with another sample of the same compound starting in the solid phase. The sample was heated at a constant rate from 15°C to 75°C. On the graph below, draw the resulting heating curve.

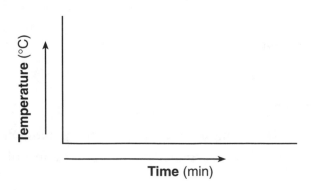

Important Formulas and Equations

1. 4   As shown on Table D, the joule (J) is the metric unit for energy, work, or quantity of heat.

2. 4   For melting or freezing, use the heat of fusion equation $q = mH_f$.

3. 4   In Table T, the equation $q = mH_f$, is used to calculate the amount of heat involved in the phase changes of melting and freezing. The heat of fusion is given in Table B as 334 J/g.

   Substituting and solving: $q = (29.95 \text{ g})(334 \text{ J/g}) = 10,003.3 \text{ J}$ or $1 \times 10^4 \text{ J}$

4. 4   The heat of vaporization is $q = mHv$. The heat of vaporization of water is given in Table B as 2,260 J/g.

   Substituting and solving: $q = (10.00 \text{ g})(2,260 \text{ J/g}) = 22,600 \text{ J}$.

5. 1   Heat always flows spontaneously from warmer objects to cooler objects.

6. 2   A change of phase occurs at constant temperature. Therefore, the average kinetic energy, measured by temperature does not change, only the potential energy changes. Line segment $\overline{BC}$ represents the time during which the sample is melting. Since melting is an endothermic process, energy is absorbed, causing the potential energy to increase.

7. 2   By definition, temperature is a measure of the average kinetic energy of the particles in a substance. The higher the temperature, the greater the average kinetic energy of the system.

8. 3   The average kinetic energy of a substance increases when there is an increase in temperature. During a phase change (choices 1 and 2), the average kinetic energy remains the same, only the potential energy changes.

9. 3   Freezing ($\ell \rightarrow$ s) is an exothermic because this phase change releases heat.

10. *a*)   Numerical setup: $q = mH_f = (25.0 \text{ g})(334 \text{ J/g})$ or 25.0(334)

    b)   Answer: 8,350 J

      Explanation: The equation, $q = mH_f$ is used to calculate the heat involved when a substance is melting or freezing. The heat of fusion is given in Table B as 334 J/g.

      Substituting and solving: $q = (25.0 \text{ g})(334 \text{ J/g}) = 8,350 \text{ J}$

11. *a*) Answer: remains the same *or* does not change

Explanation: During segment $\overline{BC}$ the substance is melting. During a phase change the temperature remains the same, thus the average kinetic energy also remains the same.

b) $\overline{DE}$ is longer than $\overline{BC}$ *or*   it takes more time to boil than to melt

Explanation: $\overline{DE}$ represents boiling and $\overline{BC}$ represents melting. Since it takes more energy to vaporize the substance than to melt the substance, the heat of vaporization must be greater than the heat of fusion. The longer line segment of $\overline{DE}$ indicates this.

12. *a*) Numerical setup: $q = mC\Delta T = (5.00 \text{ g})(4.71 \text{ J/g}\cdot\text{K})(30. \text{ K})$ *or* $(5)(4.71)(30)$
Answer: 706.5 J Significant figures need not to be shown.

Explanation: In time interval AB the liquid ammonia is heating up 30 K from 210 K to 240 K. Since there is a change in temperature the equation $q = mC\Delta T$ must be used. The known variables are: $m = 5.00$ g, $C = 4.71$ J/g•K, and $\Delta T = 30$ K.

Solution: $q = (5.00 \text{ g})(4.71 \text{ J/g}\cdot\text{K})(30 \text{ K}) = 706.5$ J

b) Answer: 6,850 J

Explanation: The equation to use for vaporization is $q = mH_v$. H_v (1,370 J/g) is given in the Physical Constants for Ammonia chart.

Substituting and solving: $q = (5.00 \text{ g})(1,370 \text{ J/g}) = 6,850$ J

Important Formulas and Equations

$$K = °C + 273$$

K = kelvin
°C = degrees Celsius

Overview:

In science, the Celsius (C) and Kelvin (K) temperature scales are used exclusively. This equation enables one to convert a temperature reading from one scale to the other. The Kelvin temperature scale is also known as the absolute temperature scale. The word degree and its symbol are not used when expressing a temperature on the kelvin scale.

Example 1:

What is the kelvin temperature that equals 56°C?

Solution: $K = °C + 273$
$K = 56°C + 273$

Answer: $K = 329$

Example 2:

What is the Celsius temperature that equals 456 K?

Solution: $K = °C + 273$
$°C = K - 273$
$°C = 456 - 273$
Answer: $°C = 183$

Additional Information:

• Absolute zero is defined as the lowest possible temperature that can exist. It is 0 K or –273°C. At this temperature, theoretically all molecular motion ceases.

• Temperature is defined as a measurement of the average kinetic energy in a system.

— Set 1 —

1. The temperature of a sample of nitrogen gas is a measure of the molecules' average

   (1) activation energy
   (2) potential energy
   (3) kinetic energy
   (4) ionization energy   1 _____

2. Which Celsius temperature is equal to 298 K?

   (1) 25   (3) 298
   (2) 273   (4) 571   2 _____

3. Which kelvin temperature is equal to 56°C?

   (1) –329 K   (3) 217 K
   (2) –217 K   (4) 329 K   3 _____

4. As the temperature of a substance decreases, the average kinetic energy of its particles

   (1) decreases
   (2) increases
   (3) remains the same   4 _____

5. What Celsius temperature is equal to 418 K?

   _____ °C

6. What is the melting point of magnesium in degrees Celsius?

   _____ °C

— Set 2 —

7. An increase in the average kinetic energy of a sample of copper atoms occurs with an increase in

   (1) concentration
   (2) temperature
   (3) pressure
   (4) volume   7 _____

8. At which temperature would atoms of a He(g) sample have the greatest average kinetic energy?

   (1) 25°C   (3) 273 K
   (2) 37°C   (4) 298 K   8 _____

9. Which Celsius temperature equals 458 K?

   (1) 185°C   (3) 458°C
   (2) 100°C   (4) 731°C   9 _____

10. Which temperature is equal to 120. K?

    (1) –153°C   (3) +293°C
    (2) –120.°C   (4) +393°C   10 _____

11. The temperature at which the solid and liquid phases of matter exist in equilibrium is called its

    (1) melting point
    (2) boiling point
    (3) heat of fusion
    (4) heat of vaporization   11 _____

12. Convert the melting point of iron metal to degrees Celsius.

    _____ °C

1. 3   By definition, temperature is a measure of the average kinetic energy of the molecules of a substance. As the temperature of a substance increases, the average kinetic energy of the molecules increases.

2. 1   In Table T, locate the Temperature equation, $K = °C + 273$.

    Solving: $°C = K – 273$
    Substituting: $°C = 298 – 273 = 25$

3. 4   In Table T, locate the Temperature equation, $K = °C + 273$.

    Substituting and solving: $K = 56°C + 273 = 329$

4. 1   The lower the temperature of a substance, the lower the average kinetic energy of its particles. See explanation for question 1.

5.   Answer: 145°C

    Explanation: In Table T, locate the Temperature equation, $K = °C + 273$.

    Substituting: $418 K = °C + 273$

    Solving: $°C = 418 – 273 = 145$

6.   Answer: 650°C

    Explanation: From Table S, the melting point of magnesium is 923 K. In Table T, locate the Temperature equation, $K = °C + 273$.

    Substituting: $923 K = °C + 273$

    Solving: $°C = 923 – 273 = 650$

NOTES

Available only at the
TOPICAL REVIEW BOOK COMPANY

CHEMISTRY MATERIAL

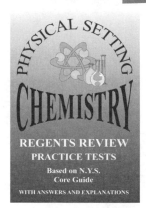

PHYSICAL SETTING CHEMISTRY
REGENTS REVIEW BOOK

Each booklet contains the 4 most recent June Regents exams, updated every September. Also includes, fully explained Answers and the Reference Tables. Each booklet is self contained with spaces for student answers.

Digest Size (8.5 x 5.5)

(ISBN 978-1-929099-20-7)   (NYC Item #901669008)

PHYSICS MATERIAL

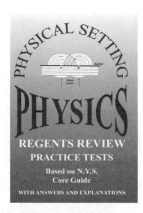

PHYSICAL SETTING PHYSICS
REGENTS REVIEW BOOK

Each booklet contains the 4 most recent June Regents exams, updated every September. Also includes, fully explained Answers and the Reference Tables. Each booklet is self contained with spaces for student answers

Digest Size (8.5 x 5.5)

(ISBN 978-1-929099-23-8) (NYC Item #929099230)

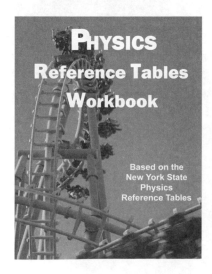

PHYSICS
REFERENCE TABLE WORKBOOK

This workbook correlates with the current NYS Physical Setting Physics Reference Tables. This workbook contains 36 sections, 26 dealing with the equations and 10 dealing with charts. Each section contains a detailed overview of the material, additional information, and a series of related practice questions. A must book for students wanting to pass the Regents.

Booklet size (8.5 x 11) (202 pages)

(ISBN 978-1-929099-87-0) (NYC Item #901668052)

WEBSITE: www.topicalrbc.com   •   E-MAIL: topicalrbc@aol.com

PRODUCING TEST REVIEW MATERIAL SINCE 1936.